D0232228

ADVANTAGE
PLAY

TECHNOLOGIES THAT CHANGED
SPORTING HISTORY

PRAISE FOR *ADVANTAGE PLAY*

'Technology has and will always be a fundamental component of sport and activity for everyone. This includes the way we capture how we do a sport and its impact on our body and health, how we book a facility or interact with friends and family, and the equipment we wear and use.

'Its growth and emergence in the past 20 years in high performance sport has provided a platform for increased awareness and consideration of the role technology now has. This book seeks to answer the big question of whether it has actually made us any wiser and improved the quality of product and the performances we now see. It provides an intriguing insight into that journey and the issues it raises for those who govern, those who promote and those who participate. If you want to understand the history and emergence of technology in sport – look no further'.

DR SCOTT DRAWER, HEAD OF TEAM SKY PERFORMANCE HUB

Advantage Play provides a fascinating and entertaining story of the role technology plays in shaping sport, told by one of the true pioneers of Sports Engineering. Guided by his passion and energy for all things science and sport, Haake charts the origins and milestones of the ways in which technology has most impacted and shaped sport, athletes and spectators. Stories are told with technical rigour and humour, unveiling the magic behinds the scenes of some of sports most famous breakthrough moments.

'Haake is a true pioneer in establishing the discipline of Sports Engineering. His passion, intellect and energy has brought scientists, governing bodies and sports brands together to revolutionize sport as we know it today. His work has impacted the breadth of athletes and spectators, from defining cutting edge coaching methods leading to gold medals for the elite, to bringing fun and engagement tools accessible to kids through programs such as sports engineering as an educational engagement tool, to the pure fun of mobile AR football tracking in the park.'

BOB KIRK, DIRECTOR, FUTURE FOOTWEAR, ADIDAS

ADVANTAGE
PLAY

TECHNOLOGIES THAT CHANGED
SPORTING HISTORY

STEVE HAAKE

First published in 2018 by

ARENA SPORT
An imprint of Birlinn Limited
West Newington House
10 Newington Road
Edinburgh
EH9 1QS

www.arenasportbooks.co.uk

ISBN: 9781909715592
eBook ISBN: 9781788851084

Every effort has been made to trace copyright holders and obtain their permission for the use of copyright material. The publisher apologises for any errors or omissions and would be grateful if notified of any corrections that should be incorporated in future reprints or editions of this book.

British Library Cataloguing-in-Publication Data
A catalogue record for this book is available on request from the British Library.

Designed and typeset by Polaris Publishing, Edinburgh

Printed in Great Britain by Clays, Elcograf S.p.A.

For The Team

CONTENTS

ACKNOWLEDGEMENTS

FIRST THINGS first: a big thank you to Rinda, Jim and Lily who have sat around the dinner table listening to my endless stories about physics and sport. Thanks for doing daft things with me and being simply wonderful. Rinda, you are a demon proof reader. And, yes, I used numerous spreadsheets in the writing of this book. *Go Team*!

Of course, I have to thank my parents for nurturing my desire for knowledge and giving me the opportunity to do so. This book was about twenty-five years in the making; twenty-four years thinking and one year writing. One obvious person to thank is Peter Tallack of Science Factory for having the confidence and insight to take my confused ramblings and turn them into something somebody would actually like to publish. While Peter 'one' introduced me to Peter 'two' (Peter Burns at Arena Sport) to make the book happen (thanks everyone), thanks also need to go to Mark Miodovnik who introduced me to Peter one in the first place and inspired me with his own book, *Stuff Matters*. Before that could happen, I met Robert Lang and Al Booth at Kensington TV from Toronto with whom I made two documentaries on science and sport. I would never have been allowed even close to those ten Olympic athletes, never mind persuaded them to try old-fashioned equipment on

TV to see how badly they'd perform. Thanks for letting me work with you and letting them appear in this book.

There are many people I've worked with in the past decades, some for over twenty years. It appears that, as a research centre, we have an accumulated experience of around 350 years: put end-to-end those years would take us back to Newton, which I think is fitting given what we do. I'd like to thank the following members of the research centre: Amanda Brothwell, Simon Goodwill, Terry Senior, David James, Nick Hamilton, David Curtis, John Hart, Ben Heller, Jon Wheat, John Kelley, Simon Choppin, Tom Allen, Leon Foster, Carole Harris, Christina King, Chris Hudson, Marcus Dunn and all the students I've ever worked with. Thanks for all the advice and support, Ben, and keeping me going on those long runs.

You might have noticed John Hart's stunning computational fluid dynamics images at the centre of the book. I feel privileged to have them in my book: thanks John.

It's surprising how much can happen during a year of writing. Small things like colds, breaking your hand; thirty or so parkruns on a Saturday morning (with thirty almond croissants afterwards). But big things too: three births and the shocking loss of two friends. Trevor and John, we'll miss you.

Getting up early every morning to write was a lot of fun, although I have to admit it was a bit tough in the dark of mid-winter, sitting in a sleeping bag wearing a big hat before the heating came on. Thank you to the BBC Radio iPlayer and 6 Music for getting me through. Each book chapter has a corresponding favourite album that I was listening to as I was writing; there is a playlist at the back of the book if you want to listen too.

Thanks one and all.

Steve

INTRODUCTION

Wanted: physicist who likes sport

IT WAS 1985 and I was looking for a job. I'd finished a physics degree at Leeds University and wondered what to do next. Margaret Thatcher and Ronald Reagan were in the middle of their special relationship, *Back to the Future* was in the cinemas, the San Francisco 49ers were in the ascendancy and Windows 1.0 had just been released, much to the amusement of Apple Macintosh users.

There were plenty of jobs for those leaving university, mostly because there were so few graduates back then. I applied furiously to the big corporations attending the jobs fairs, all of them keen to snap up the best talent. I was rejected for a Willy Wonka-style job at Cadbury's Bournville factory, but soon fell on another that looked promising: semiconductor scientist at British Telecom, helping to build a new generation of computer chips for the world's telecommunications industry. The lab was in the middle of a marsh and I arrived on a dank, misty day. I'd heard about sad buildings, but this one was seriously depressed. I turned the job down.

The next job I went for was in image processing, an opening in the emerging software sector. Personal computers were still quite primitive back then and you were lucky if you had a hard disk bigger than a few megabytes; in contrast, your phone today has

10,000 times more space. Reagan had recently launched his 'Star Wars' initiative where US orbital platforms would launch defence missiles at Russia's Intercontinental Ballistic Missiles before all-out war would probably destroy us all anyway. It was only during the job interview that I found out what it was really about: smart bombs to destroy enemy tanks. I went back to the jobs pages.

I was ecstatic when I saw an advert for a physicist who liked sport. Surely, that was me? I loved playing football, squash and tennis and enjoyed watching sport even more. It didn't say that you had to be any good, which was lucky, because I was high on effort, low on technique. The only problem was that the PhD was on the mechanics of golf. I didn't play golf and didn't really understand it. I applied for the position and found to my dismay that the other applicant was a passionate golfer with a good handicap.

Astonishingly, I was offered the post: the panel thought that my rival's love for golf might have distracted him from research. This is probably the only time in my life when being bad at something has been a bonus.

Against most people's advice, I accepted the studentship and went to work at the unlikely named Sports Turf Research Institute in Bingley, West Yorkshire, with regular visits to Aston University to see my PhD supervisor. Based on the top of a cold hill, I was thrown into a 40-strong company full of agronomists, biologists and greenkeepers and told to come back in three years with a thesis on what happened when a golf ball hit a green. I loved telling people what I did because they never really knew whether to believe me. I remember a girl asking me repeatedly at a party what I did and the only way I could shut her up was to tell a more plausible but blatant lie – that I twinned towns.

As a non-golfer, I hadn't realised how big the golf sector was. I visited the golf equipment giant Titleist in New Bedford to see their research facilities, designed to eke out ever-diminishing improvements in balls and clubs. I visited the United States Golf Association in New Jersey, with their equally impressive research centre to limit how far companies such as Titleist could go. It put me in mind of a quote by George Orwell: 'Serious sport has nothing to do with fair play... it is like war minus the shooting.'

Some three years later I emerged from Aston University with a PhD with the rather niche title of 'Apparatus and test methods for the impact of golf balls on golf greens'. I'd given golf a go during my PhD but, being six feet four inches tall, I never learnt to swing the club so that it consistently met with the ball. Any passion I might have had for the game evaporated and, instead, it turned to the physics and technology of sport. By the 1990s, I was part of a new community of researchers working feverishly on sports technology, riding the wave of cheap imports from China.

The early 1990s was a time when the concept of sports technology was still relatively unsophisticated and declaring you were a sports engineer in the academic world was greeted by an embarrassed silence. But the real world was delighted and I joined the lecture circuits giving popular science talks to professional societies and schools. My topics ranged from cricket ball swing to carbon fibre tennis rackets; I even appeared on the radio to discuss one of the goals that coined the phrase 'Bend it like Beckham'.

The focus of sports technology at the time was on materials and design, driven by the seemingly endless supply of carbon fibre: rackets, bikes, shoes, boats – everything had to have carbon fibre in it, whether it was needed or not. With this momentum, in 1996 I proudly organised the First International Conference on the Engineering of Sport, the full title written meekly inside the cover of the book of proceedings. On the outside, it merely said 'The Engineering of Sport' as I didn't know whether this first conference might also be the last. I needn't have worried, it was a roaring success and the conference went from strength to strength, being held every two years in places such as Sydney, Kyoto, San Francisco and Munich.

As I patted myself on the back for being the first at something for once in my life, I received a letter from a Professor Sadayuki Ujihashi in Japan who congratulated me on my conference and casually announced that he and his colleagues were having their seventh conference on the topic later that year. The lesson I learnt – and many times since – was that if you think you've come up with a new idea, think again, because someone has probably done it already.

Time is an illusion

A couple of decades on, the discipline has matured and I can hold my head up high as I walk down the corridors of academia; they even let me call myself Professor. But sports technology isn't always universally liked and some people claim that modern technologies in sport are tantamount to technological doping. Today, if I mention the phrase 'sports technology', everyone seems to immediately think of their smartphone or wearable rather than materials such as carbon fibre. People always ask me what's coming next, whether winning is all about the technology rather than the athlete, how we decide whether technological improvements provide a fair (and legal) advantage or are cheating – and even whether we should guard against 'technical doping' (after a hidden motor was discovered in a rider's spare bike at the 2016 Union Cycliste Internationale Cyclo-cross World Championships).

Yet to answer these sorts of questions, we need to look back in time. In any particular era, we have a certain set of sport technologies at our disposal and a set of cultural values that define what we are willing to allow. For instance, people are happy to allow polyurethane running surfaces that give world record times but aren't so happy for runners to put springs on their feet. Starting blocks, considered cheating in the 1930s, are now an essential part of sprint starts. So to my mind at least, if the rules are clear and unambiguous, and a technology is described by the rules and permitted, then it is very clearly not unfair or deceitful. Problems arise when the rules are badly written or when an unexpected technology appears that changes the nature of the sport. That's when people start to become suspicious.

Progress in sport is often so gradual that we have the illusion that the rules don't change. But pick up a vintage gazette and you'll quickly see that sports have altered hugely over the generations. The size of a ball, the number of games before a tiebreak, the way you're allowed to tackle: all these had to be decided at some point. Often, the initial choices were a best guess but then changed as either the style of play altered or new technologies emerged. Take football for example. The oldest football club in the world is Sheffield FC and their rules were published in 1858[a]. With 11 rules, the first ten were associated with the game and only the last one was related to apparel, stating:

'Each player must provide himself with a red and dark blue flannel cap, one colour to be worn by each side.'

In these first rules, the equipment doesn't get a mention, not even the ball. But just five years later, the newly founded International Football Association Board agreed a set of 14 rules including this one for footwear:

'No player shall be allowed to wear projecting nails, iron plates or gutta-percha on the soles or heels of his boots.'

In football's first games, players would generally play in their one pair of boots. Not only were they cumbersome to run in, but the smooth leather soles were treacherous on the muddy pitches of the time. It didn't take long for players to try their own makeshift studs with spikes and nails hammered into their soles. Evidently some of them were too dangerous, hence the new rule. This shows the balance between rules and technology: technologies are made to satisfy the needs of the players; if they're not liked, the rule changes and the technology changes accordingly.

---

[a] Sheffield FC is the oldest football club in the world that is independent of any school, workplace, hospital or church. Notts County is the oldest professional football club and was founded in 1862.

A hundred and fifty years on, the rules for soccer now contain 12,000 words. The ball is standardised, a regulation introduced after the first World Cup in 1930 between Argentina and Uruguay when each team brought their own ball. As a compromise, they got to play with their ball in one half each, the Argentine ball in the first half and the Uruguayan one in the second. The Argentines scored twice with their ball, the Uruguayans three times with theirs.

Goal-line technology is the thing of the moment in soccer. It uses sophisticated cameras to track the exterior of the ball as it crosses the goal line. It was resisted for many years by FIFA, soccer's international governing body, but the turning point came after the 2010 World Cup match between England and Germany in Bloemfontein, South Africa. Frank Lampard picked up the ball on the edge of the penalty area and lashed in a quick shot to beat the German goalkeeper. The ball hit the underside of the crossbar, rebounded down about a metre over the goal line, bounced back up to hit the bar once again, before rebounding into the goalkeeper's arms. The English coaches leapt to their feet, Lampard wheeled around in delight and the England fans celebrated. But the goal, which would have drawn the score level at 2-2, wasn't given. What was clear to most in the stadium as well as everyone watching on TV was completely missed by the match officials. Replays showed again and again just how bad the decision was. Even worse, England had been in the ascendancy at that moment, but went on to lose 4-1. An embarrassed FIFA sanctioned goal-line technology and wrote an appropriate set of rules for everyone to work with.

The reason that the technology is possible in the first place is the rapid evolution of digital cameras. Simply computer chips with a lens on the front, they allow anyone with the latest smartphone to take high-resolution images at 200 times per second. Aim an array of cameras at the goal line, do some image processing and, *voila*, you have goal-line technology.

A player plucked from Victorian Britain would have been surprised to find out that the technology was based on cameras; to him, taking a photograph involved standing very still in front of a large lensed box containing a glass plate and then waiting for it to be developed, printed and framed. It's the exact opposite of what you need for goal-line technology, which has to capture motion and process the images almost in real time.

Camera technology has advanced, a sporting need appeared and innovators such as Hawk-Eye have emerged – the first company to receive a FIFA licence for their goal-line technology systems. FIFA created a whole set of specifications to define what the systems should be able to do. It's difficult to say which came first, the technology or the need. Was there only clamouring for it because we knew it could be done? Or was the need actually there and technology was just the solution? I suspect it was probably a bit of both. The point is that the rules of sport evolve and developments in technology are sometimes the catalyst.

This sporting life

In this book, I tell the dramatic stories of technological breakthroughs across thousands of years of sporting history – discoveries that changed the relationship between technology and sport, the delicate balance between tradition and modernity, and sometimes even the rules of sport themselves.

The breakthroughs I've chosen might be unexpected but come from the projects I've worked on in my 30 years of experience of sports engineering: research spanning the mechanics of golf, tennis and football; the aerodynamics of sleds and rackets and balls; the traction of boots, wheels and turf; and the capture of data to improve the performance of Olympic athletes.

Together these breakthroughs reveal that the way we design sports equipment has always been much the same. Rather than being a new thing, sports technology is actually as old as civilisation; only the phrase is new.

The obvious starting place for any book on sports technology has to be ancient Greece, although there's a surprise entry from an even older civilisation from Central America. I then skip forwards through time, passing quickly through the Dark Ages which, from a sports point of view, are pretty dark. The Renaissance of the 14th and 15th centuries signified the birth of some of our early sports, but most of those we think of today appeared in the cultural and technological big bang that was Victorian Britain.

The problem I had when I organised my conferences was how to split sports technology up. Should I do it by topic? Materials, design or electronics and so on? Or should I do it by sport? Tennis, golf, football, athletics? The same problem applied to this book and I chose to go through it all in chronological order from the Greeks onwards, introducing sports as they were created and asking which technologies affected them most.

Arriving at the present day, it's probably no surprise to find that we now define sports technology in terms of the electronics we carry or wear, stimulating us with burgeoning amounts of data. But data is not enough; any new knowledge has to improve performance in some way and, hopefully, examples of my research centre's work with our Olympics teams shows how this is done.

Finally, I ask: 'Where will it all end?' If our modern Olympic Games are to last as long as the ancient ones did, then we'll still be holding them in 3036. Data shows unequivocally that performance in sport is plateauing, so how will we satisfy our insatiable desire for improvement and world records? Performance will only change if something forces it to and it's likely that technology will be one of the catalysts. What will our sports and technologies look like in the coming centuries? Will we resort to genetic modification or robotic implants? What will we consider cheating?

This book covers around 4,000 years of the past, present and future of sports technology. Behind every technology in every sport in every era there are amazing people with stories of inspiration, success and tragedy. What I've found is that sport, technology and society are intertwined – after all, technology is the way in which we humans try to improve things. It can help create a sport and it can help improve it, but occasionally it can go too far. The key is for us to decide exactly what too far means.

Steve Haake
Autumn 2018

Advantage play

n. method of gaining a legal advantage during gambling; for instance, wearing sunglasses while playing poker.

Humphrey Bogart. Hollywood liked to portray its stars as active sports people. This is a snapshot of sports technology *c.*1945. Photograph by Scotty Welbourne, © *John Kobal Foundation; Getty Images*

ONE

Starting from scratch

HE LAUNCHED himself down the track, little explosions of dust rising from his heels. I watched him from the finishing line as he relaxed into a smooth rhythm. Suddenly, he was on me: I followed him through the finishing tape as it broke and cascaded off into the wind, pressing the button on the stopwatch as hard as I could, worried I might have missed him.

He was already walking back towards me with that muscular swagger that sprinters seem to have. He raised his eyebrows questioningly;

"How'd I do?"

Andre De Grasse was just over 20 years old and already the darling of Canada. And yet, following a week in which he'd won both the 100 and 200 metres golds at the Pan American Games in his home town of Toronto, here he was helping me with an experiment on a high-school cinder track, a million miles from the screaming crowds that had witnessed his triumphs.

The experiment was part of a Canadian science documentary and when we met for our experiment at Pickering High School in the suburbs of Toronto, Andre was a little stiff after his week of triumph.[1] We were interested in the impact of technology on sport. There are two ways of working this out: you can give someone a better technology and see how much their performance improves, or, you can give them an old technology and see how much worse they get (this assumes the old technology is not as good, of course). We'd chosen the latter.

Figure 1. One of Jesse Owens' running spikes from the 1936 Berlin Olympics. Given the political circumstances of the time, it is amazing that Adi Dassler, the founder of adidas, both wanted and persuaded Owens to wear his shoes. Following his four gold medals, Owens sent them back to Dassler with a thank you note. © *adidas*.

Andre would run in a pair of shoes similar to those used by Jesse Owens at the 1936 Berlin Olympics – single-skin leather, no padding and leather soles with six long heavy steel spikes screwed into the forefoot. These were very different to the lightweight Nike sprint shoes he normally used. Teams of researchers have optimised these over the years and found that energy is lost during the bending of the toes. Increasing the sole stiffness to stop this bending can improve times by about one per cent. This might not sound like much, but it could be a tenth of a second over 100 metres, enough to win gold rather than silver.[2] Did this mean that Andre had just lost a tenth of a second simply by putting on an old pair of shoes?

De Grasse wouldn't run on a modern polyurethane track either. He would run on a high-school cinder track. No starting blocks, no electronic timing system, just me and a stopwatch. I scratched a start line across the cinders with my foot, used spray chalk to mark out a lane and measured out 100 metres to the finish where I would stretch out a tape between two poles. This is where I would wait with my 1930s stopwatch, accurate to one fifth of a second. He'd only be able to run flat out once so, as the single timekeeper, I was a little nervous.

Before starting blocks were allowed, it was customary for runners to dig holes in the ground for footholds. So, like kids on a beach, Andre and I used our hands to scoop out the cinders. This was a little concerning – I'd seen a video of Jesse Owens doing the same at the 1936 Olympics and he'd had to stab hard at the ground with a trowel. Perhaps this track was a bit soft?

The starter in Berlin had used a starting pistol but Canadian law didn't allow us to use one in a residential area. Andre's coach had it all figured out – he would slap together two smooth wooden blocks to make a sharp retort like a gun.

I walked to the finish to look back down the rough track; only then did I realise we had a problem. Back in 1936, the timing judges would have used the smoke of the pistol as a cue to press the button. I wouldn't be able to see the wooden blocks moving from 100 metres away and it would take three tenths of a second for the sound to reach me. We hastily arranged for a production assistant to stand by the starter with her arm in the air. She would drop it when the coach slapped the blocks together.

I saw Andre kneel down on the start line and the production assistant raise her hand. Her arm dropped and Andre was away. I squeezed the button to start the watch and I heard the faint slap of the blocks finally reach me. Forty-seven dust-raising steps later, Andre flashed through the finishing tape.

He sauntered back to me, panting.

I squinted at the watch. "Eleven seconds," I said.

"I think," I murmured nervously.

Reaction

It's surprisingly difficult to time a race. Did I press too early? Did I press too late? It all happened so fast, I couldn't really tell.

One of my PhD students – Leon Foster – had written his thesis on the effect of technology on elite sport with the rationale that if technology affected sport, then it would appear in the results.[3] Over a three-year period, he collected, checked and cross-checked more than 60,000 results going back to 1891 for all men's and women's track and field events.

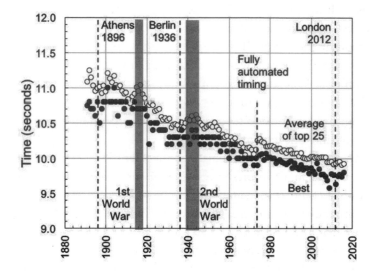

Figure 2. Performance in the men's 100-metre sprint from 1891 onwards. The upper line of circles represents the mean of the top 25 for each year while the lower dark circles are the best performances. The current world record is 9.58 seconds set by Usain Bolt in 2009; the transition from manual stopwatches to fully automated electronic timing, created in 1975, is clearly seen as a step change. *Data courtesy of Leon Foster.*

His results for the men's 100 metres explained something surprising about stopwatch timing. Times dropped from an average of 11 seconds in 1891 to less than ten seconds today. With obvious improvements in nutrition, coaching and professionalism in sport, this was something we'd expected. But then we found a big step change around 1975 which coincided with the change from manual stopwatches to fully automated electronic timing.

But this is the weird thing: the athletes' times had got worse.

Almost overnight, average times for the top 25 men increased by two tenths of a second from around 10.1 to 10.3 seconds; it was as if they had an extra two metres to run.

With modern fully automated timing, the measurement starts with the bang of the gun and finishes with the athlete crossing a light beam at the finish. The timing is accurate to one hundredth of a second. With stopwatch timing, the measurement starts and ends with the pressing of the watch's button. The first press is the trickiest to do because it takes time for the brain to process the reaction at the hands. This takes about two tenths of a second, precisely the jump we saw in the data. The International Association of Athletics Federations now recommends that 0.24 seconds be added to stopwatch times when comparing to fully automated timing; how they came to this number appears to be lost in the mists of time but it seems about right.[4]

I checked this for myself by analysing one of Jesse Owens' heats from the 1936 Olympics[5] in which his official time was 10.3 seconds. I counted the video frames of his race from the first puff of gun smoke to him crossing the line; there were 319 frames. Since the film was recorded at 30 frames per second, this gave the real time as 10.63 seconds, more than three tenths of a second longer than the official time, artificially reduced by the timing judge's slow thumb.[6]

Back at the finishing line in Toronto, Andre De Grasse waited patiently for me to give him his time. I wondered what to say. I needed to add the IAAF amount to the time on the stopwatch – this would give 11.24 seconds. I felt like I was giving him bad news and I didn't want to tell him. He'd run 10.07 seconds in the Pan American Games final just a few days before; how could I tell him he was over a second slower?

I looked for an excuse. He could just have been tired, of course, and this was hardly the excitement of a packed stadium so he wouldn't have the energy boost of the crowd to spur him on. The starting blocks, or lack of them, would surely have made a difference as would the track itself which, at the Pan American

Games, was made of deep-red polyurethane by a company called Mondo. Polyurethane tracks first came to the world's attention in the 1968 Olympics in Mexico under the brand name Tartan. Ten world records were broken in the 100, 200, 400 and 800 metres events. Most commentators put the records down to the thin high-altitude air of Mexico City but many still remember it as the moment the Tartan track made its international debut.

Ever since Mexico, companies like Mondo have claimed that their tracks are optimised for world records. Was this just advertising hype or was there any truth to it? I picked up one of their brochures and read their explanation:

> During each step, the force gained and not dissipated transitions from the 5th metatarsus to the 1st metatarsus faster with Mondo than with existing track products.

Don't worry if you don't understand what this means – I don't either: I think the science has been mangled by someone in marketing. Two people who could easily explain it, however, are Thomas McMahon and Peter Greene of Harvard University who did research in the 1970s to see if there was anything to the new tracks.[7] They created a mathematical model of the lower leg, simulating it as a spring and damper attached to a rack and pinion, a bit like a car's suspension. It looked like a bad attempt at a robot.

They found out that if the surface was too soft, it absorbed energy that the athlete could never recover. If the surface was too hard, then the leg muscles would have to work overtime to slow down the short, sharp impacts to manage the high loads. There was an optimum in the middle where the deformation of the impact was shared equally by the surface and the lower leg. An optimised track effectively kept the foot contact time short enough so that the athlete spent more time flying through the air, but long enough so it wouldn't cause injury.

I looked at videos of Andre's 100-metre final at the Pan American Games and compared it to his run on the cinder track. Surprisingly, he took the same number of steps in each – 47 –

making his average stride length a little over two metres. If he took the same number of steps but ran faster on the polyurethane, then he must have turned his legs over at a higher frequency. On the polyurethane he ran at 4.7 strides per second; on the cinder track he managed only 4.2 strides per second.

This was the clue to the difference between the tracks. On the soft cinders, Andre adapted by stiffening up his leg muscles to increase the contact forces and reduce the contact time. But there was only so much he could do as the surface physically shifted and sprayed into the air as he ran. Not only did this take energy, it took time.[8] The consequence for Andre was that the soft cinders increased his foot contact time by about 25 per cent, adding up over 47 steps to more than a second.

"I don't think I've ever felt so tired running 100 metres as I do right now," he said.

Toeing the line

Using 1930s technologies, Andre's performance worsened by over ten per cent. Before every Olympics, there are the inevitable TV montages of past athletes like Jesse Owens interspersed with terracotta images of muscular naked runners from ancient Greece. What sports technologies did they use thousands of years ago? Did they even have any?

I went to the birthplace of the Olympics at Olympia to find out. It's nestled off an inauspicious side road next to a slowly meandering river on the western side of the Peloponnese in western Greece. It's hard to imagine the roar of the 40,000-strong crowd that visited every four years over 2,000 years ago. The stadium itself is merely an elongated shallow bowl with an oblong track of hardened mud. An entrance tunnel used to lead from the temple complex into the stadium; all that remains is a single arch of bricks.

I walked out onto the track and found what I was looking for – the starting line, still here after all these years. At some point, scratching a line in the dirt had become inadequate; instead, they'd put a marble sill across the full width of the track. The sill, a bit like a long kerbstone, had two parallel grooves about five centimetres apart along its length. The runners would stand on this and place their toes in the grooves, left toes in the front one, right toes in the rear with their arms outstretched in front of them in a pose more like a dive than a run. There was a second sill at the other end of the track at a distance of 600 ancient Greek feet, or about 192 metres.

The short sprint of one length of the track was called the *stade* and started at the far end of the track so that athletes ran back towards the temple complex, with the finishing line marked by posts on the sill I was standing on. The longer sprint was called the *diaulos* and was two lengths of the track; the runners would start where I now stood, run down their lane, around a vertical pole stuck in the sill at the other end and back down the adjacent lane.

The first recorded winner of the sprint at the Olympics of 776 BC was a baker called Koroibos.[9] Vases from the period show that Koroibos would have run naked. The reason for this has two explanations: one story has it that in an earlier race, runners tripped over someone's accidentally discarded loincloth, leading to a subsequent ban on all clothing on the grounds of safety; an alternative story has it that, since women weren't allowed to compete, running naked was a way of ensuring there were only men in the race.

There are images of naked athletes covering themselves in olive oil in the *gymnasion*, a word derived from *gymnós* – Greek for nude. The naked athletes were slathered in olive oil for a reason not entirely explained, although there are a variety of good suggestions: it made the athletes feel and look good; it gave them a tan; it enhanced muscle definition; it reduced dehydration from the hot Greek sun. One thing is certain, it added to the spectacle.

An essential piece of gym kit was a small bottle called an *aryballos* which held an athlete's own personal stash of olive oil. Also in their kitbag was a curved hand tool called a *strigil* which was used to scrape off the oil after a workout. The resulting mixture of sweat oil and dust sounds pretty disgusting to us today but was highly prized in ancient Greece. With an appropriate-sounding name – *gloios* – it was possibly the first athlete-sponsored product ever sold, with claims that it was good for inflammation.[10] It was presumably sold with the warning 'for external use only'.

Standing on the line in the quiet, ancient Olympic stadium, I was moved by the connection with the past. I imagined what it must have been like as an athlete representing your city in front of every other Greek city state in the civilised world, walking naked through the tunnel into the hot stadium. My reverie was broken by a group of chattering Japanese tourists anxious to take turns having their photographs on the starting line. Someone had even brought a crown of leaves for an impromptu race, and they bounded off, cameras bouncing around their necks as they ran. Everyone loves a race.

The Olympics wasn't the only event in ancient Greece: there were three other 'Crown Games' at Delphi, Isthmia and Nemea. I bought a book called *Ancient Greek Athletics* and my jaw dropped. There were photographs of a start line with a catapult-driven starting gate called a *hysplex* and people in white tunics racing down the track.[11] I quickly checked the caption: the 'Nemean Games'. A little more research revealed that the Nemean Games were recreated every four years, during Olympic year. I could actually take part and see what it was like to run as they'd done in ancient Greece. I even found to my delight that I could volunteer to help set up the starting gate with the man who'd unearthed and reconstructed the stadium, Professor Stephen Miller.

A 2,000-year-old puzzle

Stephen Miller, from the University of California at Berkeley, was given the job of excavating Nemea in 1971. The Nemean stadium complex was minimal at the time with only a single column of the Temple of Zeus and a vague indentation in a nearby hillside where a stadium was supposed to be. By 1974, he'd inspired enough benefactors to collect money to buy the land and begin to excavate. He wasn't universally welcomed and a local fired a gun in his general direction as a warning. But there was support from the Mayor of Nemea who gave one proviso: he would support Miller, but only if the Nemean Games were recreated every four years once the stadium was unearthed. Miller agreed.

They dug out six metres of soil in the first four months of the summer of 1974 but found nothing. Money to pay the workers started to run out but, on the last day of digging, in a hole seven metres deep, they hit stone: it was the marble starting line. It took another 20 years of digging to unearth the whole sanctuary complex, including the baths, locker room, water channels and basins. But most important of all, they uncovered the 13-lane running track. Originally 600 ancient Greek feet long (about 192 metres), the end of the Nemean track had been lopped off by farming and was now only 90 metres long.

Forty-seven years on from that initial dig, I walked past the road to the museum, now called ΟΔΟΣ ΣΤΕΦΑΝΟΥ ΜΙΛΛΕΡ, Greek for 'Stephen Miller Road'. I found Professor Miller on the track arguing in Greek with a small cohort of ageing helpers as they brought out the components for the starting gate. They were animated and pointing at the blocks of wood by their feet, evidently trying to work out how to put them together.

I tentatively offered my assistance and was given the important task of fetching, carrying and holding. This gave me an opportunity to watch them build up the *hysplex* starting gate which was retrofitted to the original stone starting sill. Vertical lane posts stuck out of the sill about a metre apart. The square holes into which the posts were pushed were originally lead-lined and the posts wetted before they were slotted in. As the water soaked into the wood, they expanded against the lead, keeping them secure (an 'interference fit' in modern engineering parlance).

The catapult was invented around the fourth century BC as a weapon of war and was adopted almost immediately as a mechanism to drive the starting gates. The torsional catapult was especially efficient and used the energy-storing capabilities of twisted sinew and hair. The purpose of the starting gate was to hold two cords across the starting line, one at waist height and one at knee height. Releasing a trigger would slam the ropes to the floor in front of the runners, allowing them to step over and begin their race.

Figure 3. The reconstructed starting gate in Nemea in action at 'ready' (left) and 'go' (right) with the gate arms slamming to the floor. The starter is just out of shot to the right and pulls the trigger ropes attached to the thick columns. The man in black robes standing behind the runners is the judge: he holds a stick to whip any false starters. © *Steve Haake*.

The construction process was more complicated than I'd imagined.[12] Everything was wedged in place using wooden blocks cut precisely to match the slots in the stone sill. This was the cause of the argument between Miller and his colleagues because no slot in the stone was identical and every piece of wood was unique. It was like putting together one of those three-dimensional puzzles you get in Christmas crackers but without the solution. In the third century BC, a Corinthian engineer called Philon was engaged to build a similar *hysplex*; evidently, he must've lost the instructions too, because it didn't work. The organisers were furious and fined him 500 drachmae, the equivalent of a year and a half's wages.[13]

After an hour or so, the *hysplex* had been made. A trigger rope led from the anchor posts at each side of the track to the starter standing behind so that the ropes made a large V. He shouted *poda para poda* (foot by foot), *ettime* (ready), *apete* (go) and gave the trigger ropes a yank. The catapult slammed forwards to the ground, taking the horizontal cords with them. Safely pinned to the ground, the runners would be able to leap over the cords and race down the track. Anyone caught up in them would be flogged by a judge in black robes for making a false start.

The track was swept clear of pebbles and 12 of the 13 lanes were marked out with lines of white lime, much like I'd done with Andre back in Toronto. Each line was marked with a single letter of the Greek alphabet. The stage was set for the next day's Games and I would be one of the 1,200 or so competitors taking part in heats of the shortened *stade* – the 90-metre sprint.

When my daughter had first found out about my intention to run in the Nemean Games, she'd looked at me in horror and asked:

"You're not running *naked*, are you?"

Luckily for all concerned, runners were given a *chiton* to wear – a white cotton tunic with a belt around the waist. We were still expected to run barefoot, though, and this was my major worry. As a pretty committed runner, I was used to wearing modern trainers with a decent amount of cushioning. As a committed amateur, however, I tended to land on my heels.

Measurements of foot impact forces during running show that most runners brake slightly when they first hit the ground; this is exacerbated by landing on the heels. As the jogging boom hit in the 1970s, the number of running injuries began to rise: the knees, Achilles tendons, shins and arch of the foot were all at risk. Manufacturers began to introduce cushioning into their insoles made from ethylene vinyl acetate foams (EVA). In sprinting, a stiffer sole was better, but for anything else it seemed that cushioning was what we wanted. Is this right? Is it as simple as that?

Unfortunately, the jury is still out on this one. Some studies have shown cushioning to improve performance, some to make it worse. Benno Nigg from the University of Calgary, one of the world's experts in running, has said that any improvements are specific to the athlete so you can't generalise. One thing that people seem to agree on, however, is that performance can be improved if you reduce the unhelpful braking forces during impact.

One way to do this is to train yourself to land on your forefoot rather than your heel. The 'pose' method of running was trademarked by Nicholas Romanov: he advocated leaning forwards so that you seem to fall forwards onto your toes.[14] Running barefoot has a similar effect. Brigit De Wit and researchers from Ghent University in Belgium tested nine male long-distance runners in both running shoes and barefoot. They found that the shoeless runners landed flatter on the foot in an attempt to eliminate the high forces under the heel which caused pain during impact.[15]

Cue the world of barefoot running. Amazingly, manufacturers have done a bit of an 'emperor's new clothes' trick and managed to turn wearing nothing into wearing something and make barefoot running a billion-dollar industry. Undoubtedly, this was helped by Christopher McDougall's best-selling book *Born to Run* which documented the feats of the Mexican Tarahumara tribe who could run long distances in minimalist sandals without injury.

Manufacturers latched on to this idea, fuelled by the worry in some minds that too much cushioning was bad, and Vibram's Fivefingers became one of the world's best-selling barefoot shoes. They are effectively rubber socks that provide protection from the surface and whatever is on it (stones, litter, dog mess) but with little in the way of shock absorbency. Although there was plenty of anecdotal evidence of their success, there was an apparent lack of scientific evidence. Vibram subsequently lost a $3.75 million lawsuit launched by 150,000 customers who disagreed with the manufacturer's claims, or possibly just wanted their money back when they didn't get any faster.[16]

Worried about my own actual barefoot running to come, I decided to get in a bit of practice on the hard, dusty track of Nemea and give myself a little bit of home advantage over my as-yet-unknown adversaries. I found myself running on my forefoot as expected, with the occasional stab of pain when I hit a pebble hidden in the soil. Despite having run many races in my time –

albeit at much longer distances – I was nervous. I'd helped set up the starting gates so I knew how they worked; I'd tried out the track; I'd had a go at barefoot running. What could possibly go wrong?

Ready, steady, er?

The rain overnight was torrential. The bolt lightning was fierce, as if the mighty Zeus was angry with us for putting on the Nemean Games without emailing him first. I awoke to a brilliant blue sky with a stadium track covered in debris. It was no longer a dry dust bowl but was smooth and dark brown with occasional deep bruises of wetness. The contestants started to arrive and gabbled impatiently outside the stadium.

After a two-hour delay, the races began, starting with the oldest athletes, some in their seventies. I was in race number 13 with eight other 52-year-old men. I stripped down to my pants in the canvas changing room and copied everyone else by slapping on a bit of olive oil for effect, although the only effect I could see was that I was now a bit sticky. I put on my white *chiton* and joined a line of nervous, balding competitors. We were led by a herald around a blind turn to an archway leading into the athlete's tunnel.

I think this perfectly arched construction is the *coup de grâce* of Miller's work at Nemea. He'd found the 36-metre-long tunnel just a couple of years after the starting sill, blocked at both ends and partially silted up. Entering the dark passage raised the hairs on the back of my neck. It seemed to take a couple of thousand years to walk to the other end and we stopped just before the exit, waiting to be announced. To my left, at about head height, I could just make out graffiti scratched into the wall. This wasn't recent graffiti, it was over 2,000 years old, left by ancient Greek athletes who'd waited to compete exactly as I was doing now.

"*Niko*" ("I win"), proclaimed one piece of graffiti.

"Akrotatos is cute", read another.

"To the guy who wrote it", responded a rival.

I'd never really thought of the players' tunnel as a key design feature before. This one was around a blind corner from the changing rooms so that the athletes would come upon it suddenly as if it was a tunnel into the afterlife. It was long enough to easily accommodate 13 athletes and their coaches, and it pulled up short of the track so that the athletes emerged into a gap in the terracing. This allowed the crowd to shout down to their heroes as they appeared beneath them, running out into the bright light.

I couldn't resist touching the graffiti for good luck. Our names were called out in order and we ran out to the cheering crowd. Just a bit of fun, I thought. But I was nervous: would I false start, would I trip over the cords and get a whipping by the judges? What was the damp track like to run on? Would I embarrass my family sitting in the stands and fall over in the funny tunic? Then again, I might even win and get the prestigious garland of wild celery given to victors.

At the starting line, we drew lots for our lanes from a bronze helmet. I drew Δ for delta, the fourth letter of the Greek alphabet and got lane four. Glancing nervously down the track, I noticed that there were still dark patches in my lane – I'd have to be careful.

I got into position and placed my toes in the grooves of the starting line; left toes in the front, right toes in the rear. The catapult arms to the left and right of the sill were pulled upwards and the starting cords rose across us. The trigger was set and the trigger ropes stretched back behind us to the judge. Then came the shout, not quite as loud as I was expecting *poda para poda… ettime… apete!*

I heard *poda para poda*, then *ettime* but didn't quite hear *apete!* The gate slammed down to the ground and everyone darted forwards while I stood there like a numpty trying to figure out what was going on. By the time I jumped across the starting cords, everyone else was a couple of paces ahead. I heard my family cheering me in the stand and sprinted as hard as I could. As I accelerated I realised that the ground wasn't as hard as it had been the previous day, but instead felt beautifully soft and smooth. I focused on running on my forefoot, gaining on the pack of runners. By 60 metres I'd caught them; by 80 metres I was just behind the last two runners to my right. I was at my limit, but I knew I could win if I just pushed a bit harder...

I've already mentioned two of the features of running that athletes and coaches take notice of – stride length and stride frequency. You might have long legs and a good stride length (like me), but a good sprinter will be able to turn their legs over fast. Both of these together give you good sprint speed.

But doing this requires good strength and conditioning of the legs, particularly the hamstrings and gluteal muscles. My problem at this point in the race was that I was at the limit and my strides were *not* going to come any faster. My body or my brain – I don't know which – did the only thing it could and started to increase my stride length. The effect was that I started to land on my heels rather than my forefoot, reverting to type. This was, literally, my downfall. As my left foot hit a particularly wet bit of soft ground, my heel slipped forwards and the muscle in my left hamstring pinged like a broken guitar string.

I limped the last four agonising strides past the finish posts to come in third.

Fittingly, the word agony comes from *agon*, Greek for contest.[17]

Back to the future

At the first modern Olympic Games in Athens in 1896, we turned to the ancient Greeks for inspiration. The stadium in Athens was the newly reconstructed Panathenaic Stadium, around the same length as the stadium at Olympia. But rather than a single straight, they had two straights connected by a tight turn at each end to give a squashed oval, 203 metres long and 33 metres wide. The ancient Greeks held long races by putting a pole at the end of the track to run around and back, but the modern continuous oval was a much neater solution.

The sprinters at the first modern Olympic Games adopted a variety of starting poses from something very much like the ancient Greeks' *poda para poda* to the modern crouching style we know today. Thomas Curtis was one of the American athletes who took the tortuous two-week voyage across the Atlantic to compete in Athens. "The track, by the way," he said, "was well intended and well built, but it was soft, which accounted in part for the slow times recorded." The winner of the first Olympic 100 metres sprint was American Thomas Burke, who ran a time of 12.0 seconds, over a second slower than the fastest that year of 10.8 seconds.

Figure 4. The start of the 100 metres at the first modern Olympic Games in Athens in 1896. Athletes adopted a variety of starting poses with the crouching athlete second from left, Thomas Burke, the eventual winner.

By the time Jesse Owens ran in the Berlin Olympics in 1936, the 400-metre oval track had expanded widthways with an open space in the middle for field events. Starting blocks had begun to appear in the late 1920s but weren't officially sanctioned by the IAAF until after Berlin in 1937. Albert Barron was one of the first inventors; he devised something that looked remarkably like the stand for an old-fashioned deckchair. His patent of 1931 described exuberantly why he thought starting blocks were so important:

> Starting blocks… are devised with a view towards obviating a stumbling and uncertain start on a running track and to afford a splendid and dependable means to facilitate a vigorous spurt from the starting point and enable the runner to literally spring or bound forth into immediate and energetic motion.

These early starting blocks were no more than simple mechanical devices to push off against. Placed behind the starting line, Barron's patented blocks had two main features: they had anchors to stop them moving and they were adjustable for athletes of different physiques. When automated timing was eventually introduced, the blocks were also used to measure false starts. These used strain gauges behind the footplates to measure any movement to within a tenth of a second of the gun, deemed as the minimum reaction time possible. A false start was punished by disqualification – at least it wasn't a flogging.

With a more accurate timing system, it became evident that the distance of the gun from the athlete would become important since the person furthest away from the pistol would hear the bang last. This could make a difference of two hundredths of a second, enough to miss out on a medal.[18] To counter this, the gun in modern systems isn't real: instead it is merely a switch that triggers a digital sound in loudspeakers placed directly behind each athlete. This means they all hear the bang at the same time and nobody gets an advantage.

Intuitively, it feels that starting blocks ought to improve performance. Researchers from the Free University in Amsterdam showed that roughly half of a good sprint time was due to the blocks and half due to the run itself.[19] I feel a little bad about Andre De Grasse, back on the cinder track in Toronto. The soft holes we'd dug for him would have made a poor version of a starting

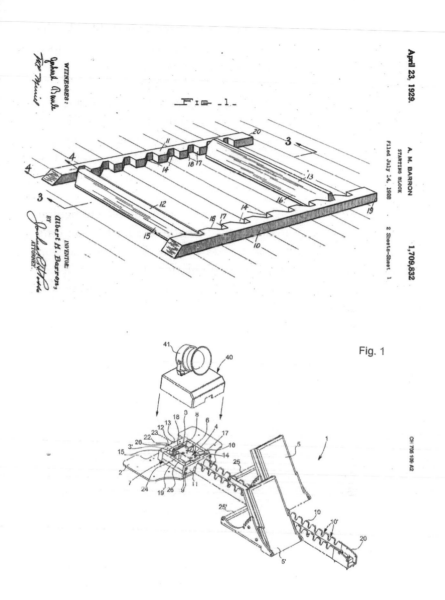

Figure 5. Patents for starting blocks. Top: 1929 patent awarded to Albert Barron from Philadelphia. Bottom: modern starting blocks from Swiss Timing including sensors to measure reaction time and false starts.

block. He might even have been better off standing. This is how he'd won his first ever 100-metre race, in a time of 10.9 seconds while wearing rather unconventional basketball shorts.

At the last Olympics in Rio, wearing proper running gear and with modern starting blocks, a polyurethane track and stiff Nike running shoes, Andre won a bronze medal in a time of 9.91 seconds. For Andre at least, it seems that all these technologies put together, along with years of training, make a difference of about one second.

Going back to ancient Greece showed me that all you really need for running is a flat piece of earth with a measured distance; you don't even need shoes, although I much prefer running in my nice cushioned ones. I became obsessed with the starting blocks at Olympia and Nemea for a while, I think because, for me, the *stade* of 600 feet defined the beginning of sport. The distance defined the size of the track and the size of the stadium, which is named after the *stade*. This simple race was so important to the Greeks that the ancient Greek calendar used the name of the winner to identify the year. If this method was still used today, the date of the Rio Olympics wouldn't be 2016, it would be the year that Bolt won his third Olympics.

The *hysplex* starting gate attached to the starting blocks was probably the first sports technology and foreshadowed our current desire for fairness. But the ancient Greeks went further than just running, of course, and the next chapter introduces the events that acted as inspiration for our modern Olympic Games; we find the world's earliest sports industry and what is probably the first personalised sports technology ever.

TWO

The shape of things to come

MONDAY 7 August 1995; Gothenburg, Sweden. A north-west wind cooled the spectators in the Ullevi Stadium where the 5th International Association of Athletics Federations' World Championships were being held. A little before six in the evening, British athlete Jonathan Edwards set off down the runway for his first triple jump of the final. Too fast to see, he hit the take-off board with his left foot and did a single six-metre hop, followed it with a massive step onto his right foot before launching himself into the air at just over 32 kilometres per hour. He leapt out of the sand with excitement, knowing he'd done something special. The white flag went up to indicate he hadn't overstepped the take-off board and he waited anxiously while they measured the jump.

The crowd roared and Edwards searched desperately for a screen: his eyes widened and his jaw dropped as he saw what he'd done. His distance was 18.16 metres (59 feet 7 inches) and a new world record. Twenty minutes later, Edwards took to the runway again. He gazed down the track at the landing pit with a smile on his face, wagging his finger at the track as if to say, 'I have you.' He sprinted down the track and completed a jump that was even more relaxed than the first. Edwards bounded out of the sand with his arms already in the air, knowing he'd done it again. He shook his head slightly as if to say, 'You wait for one world record and then two come along at once.' He'd jumped 18.29 m (60 feet), a world record that still stands today.

The triple jump is a pretty specialised event. While the long jump, high jump and pole vault have just one phase, successful triple jumpers must complete three take-offs and landings one after the other. A good jump is the summation of these three parts, all happening within just a couple of seconds.

But why have three jumps? Jumping over things is something everybody has done at some point in their lives, perhaps in the garden or on the beach. Admittedly, the pole vault is just as weird as the triple jump, but at least it was related to the feat of dyke-jumping in the Netherlands. But a hop, a skip and a jump? If you had to design an athletics field event, this wouldn't be the first idea you'd come up with. Why is it in the Olympic Games? Why does it even exist?

A weighty matter

There was no triple jump, pole vault or high jump in ancient Greece, but there was the long jump. It formed part of the pentathlon along with the sprint, javelin, discus and wrestling. The ancient long jump appears to have raised more heated discussion in scholarly circles than any other sport.[20] The reason for this is that pictures from ancient Greece show athletes using jumping weights called *halteres*.

My questions are the same as all those before me: Why would you carry weights during a long jump? Were they a mandatory part of the sport? Did they help or hinder?

The success of the ancient Olympic Games inspired other cities to create their own events. The four games at Olympia, Nemea, Isthmia and Delphi were called Crown Games because the winner's prize was a symbolic crown of leaves rather than a direct prize. Don't think that winners were hard done by, though, as they returned home to lifetime pensions, free meals, theatre tickets and cult status. They could even go professional and compete at the many Money Games that sprang up where they won cash or something equally valuable such as olive oil. Successful athletes could become very rich indeed.

The prevalence of ancient festivals means that there is a fair amount of evidence lying around to help us solve the puzzle of the ancient Greek long jump. Firstly, there are the images painted onto vases, cups and amphorae (tall jugs with two handles for holding oil and wine). Secondly, there are the jumping weights themselves; and lastly, we have a few texts which report retrospectively the amazing feats of the athletes.

Pictures show athletes in a variety of poses: training, jumping, taking off, flying through the air, landing on the ground. Successive historians have rearranged the images like the individual frames of a lost film to tell the story of a single jump. They think that the athlete started by leaning back on his right leg (it was always a man, never a woman), with his left leg forward and arms stretched out in front, weights in hand. This pose is thought to have been a crucial concentration phase, rocking back and forth in time to accompanying flute music. He then took a short run-up to a take-off board, swinging the weights backwards and forwards a couple of times in unison just before take-off. He would then throw his arms forwards at the same time as he launched himself up into the air. While in flight, he used his stomach muscles to pull his arms downwards and his outstretched legs upwards before landing on both feet. Some historians have suggested that the athlete might have gained a few extra centimetres by throwing the *halteres* away behind himself.

The weights varied in mass between one and two kilograms and were made from stone, lead or bronze. The earliest example of a *halter* is a heavy flat lead bar around about the size of a small smartphone and with a slight narrowing for the grip. A very different style was a curved stone *halter* with a hole to allow it to be grasped tightly with the fingers. There was also a flat metal *halter*, curved around the heel of the hand and with the majority of the mass in a square or wedge shape towards the front. Finally, a rather

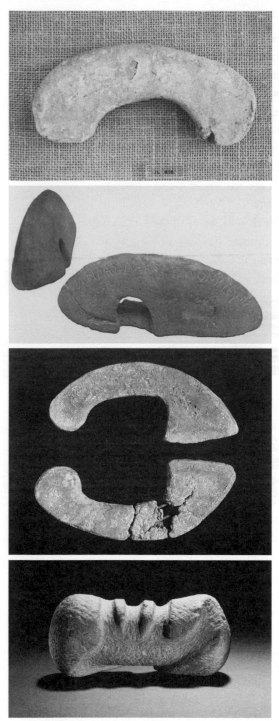

Fifth century BC. Lead, Nemea. © *Steve Haake*.

Fifth century BC. Stone. 4,629 grams. Dedicated to Zeus by Akmatidas of Sparta, probably not used for competition. *Archaeological Museum of Olympia.*

Possibly fifth century BC. Lead. 1,070 grams. © *British Museum*.

Roman. Left hand stone jumping weight, with carved finger grips. 2,230 grams. © *British Museum*.

Figure 6. Different designs of jumping weights (*halteres*) from ancient Greece.

intricate version seen on later Roman statues, was a cylindrical stone bar with grooves carved into it for the fingers.

A lot of effort was put into the grip to make sure that the athlete could hold on to the weights. It seems unlikely to me that the athletes would throw them away at the end of the jump; the weights are just not designed to get rid of that easily. Some have suggested that the *halteres* were used to make the long jump difficult, but then I would expect them to be standardised for each tournament and, anyway, at least one inscription says the athlete won the long jump 'because of this *halter*'. What I find fascinating is that the designs evolved in complexity but were all used at the same events, similar to the way sports equipment is used today. It seems likely to me that the athletes and the artisans making the *halteres* were using the designs to improve performance.

But how did they work?

Let's first consider what the Greeks understood about running. The 600-foot *stade* wasn't the same length at each venue as it depended whose foot was used to do the measurement: the track at Halieis was only 166.5 metres long, while the one at Olympia was 192 metres.[21]

The *diaulos* (twice the *stade*) appeared 52 years after the first Olympics in 724 BC, followed closely behind by a long-distance race of 20 laps or more, the equivalent of our 5,000 metres. At Corinth, the stadium had numbered lanes for runners to start in until they reached a break line at 200 feet indicated by a post. After this, they merged into a group and aimed for a single turning post at the end of the stadium, going around this and returning en masse. Further laps required the runners to go around turning posts at both ends.

Negotiating the tight bend around the post – called a *kampter* – must have been quite chaotic in the early laps when the runners were bunched together. Images of these endurance runners show them with their arms bent at around 90 degrees and their arms horizontal and close to their chests. Modern distance runners run in the same way, moving their arms as little as possible to conserve energy. Sprinters, on the other hand, run with their arms extended to increase the push-off from their feet.[22] This is clearly shown on vases, although we have to be a bit cautious about artistic licence. Sprinters are sometimes shown with the same arm and leg forward: this improves the aesthetic of the picture as it opens up the body to the viewer, but is an impossible way of running (try it!).

If the ancient Greeks understood the different styles of running, how did the long jumpers run? They looked more like sprinters than endurance runners but they certainly didn't look like they were going at full speed. The godfather of the study of ancient Greek athletics, E. Norman Gardiner, suggested that they had a few short springy steps, holding the weights by their sides. Judith Swaddling from the British Museum pointed out that the take-off point was only 60 feet from the edge of the track so wouldn't have allowed much of a run-up. It seems, then, that the run-up might have been more about coordination rather than speed, somewhere between the style of a sprinter and that of an endurance runner.

But why jump with weights? The answer seems to lie in the concept of 'inertia': the more massive an object, the more inertia it has. This means that it is less likely to start moving but, once it's moving, is also more difficult to stop. An athlete running holding a pair of weights, then, has more inertia than one without. But the athlete also swings the weights at the jump and this brings in an associated feature called the 'moment of inertia': this inertia applies to objects when they are rotating. The higher the moment of inertia, the harder it is to get rotating but also the harder it is to stop it once it's moving.

The classic example of moment of inertia is that of a figure skater spinning on the spot with arms outstretched. With the arms away from the body, the moment of inertia is high and the skater spins slowly; when the skater brings the arms inwards, the moment of inertia reduces by almost a half and the skater spins more rapidly. This shows that while the moment of inertia is proportional to mass, it is also proportional to the distance the mass is from the axis of rotation. In fact, it is proportional to the square of the distance which is why the skater's spin increases so dramatically when the arms are brought inwards.

If you were an athlete back in ancient Greece, how would you choose the best pair of *halteres*? To show you how they might have done it, let me introduce you to an ancient Greek pentathlete called Phayllos from fifth century BC Kroton, a Greek colony in southeast Italy. After winning the pentathlon in Delphi in 482 BC, he bought, crewed and commanded a ship for the Greeks at the Battle of Salamis where they unexpectedly beat the larger Persian army. He returned to Delphi and won the pentathlon again in 478 BC. To the ancient Greeks, Phayllos was both a hero and a superstar, a mix between David Beckham and Winston Churchill.

In an *emporion*, Phayllos might have found the *halteres* arranged by weight, material and price. I'd expect stone weights to be the cheapest and metal weights to be the most expensive. Phayllos' selection process would have been similar to the one we use today when selecting, say, a tennis racket in a store. I once knew a tennis-loving physicist called Howard Brody and he told me that there were three things that players looked for when selecting a racket.[23] The first was the right mass – this was the first thing the player perceived when picking it up. The second thing was the mass distribution: was it head heavy or head light? This was felt when the racket was balanced in the hand. The third was the moment of inertia (or swingweight as tennis players prefer to call it). This was gauged when the racket was swung through the air: a high moment of inertia would make it difficult to swing and vice versa.

Phayllos would probably have gone through the same process when selecting his *halteres*. He would have picked up one that had the right mass, the right balance in the hand and that could be swung easily in the air. Knowing the physique of Phayllos allows me to work backwards to find out what he might have chosen. Research on skeletons from the period estimate that Phayllos would have weighed around 70 kilograms and been about 170 centimetres tall.[24] If the *halteres* were chosen to roughly double the moment of inertia of his arms – as the heaviest tennis rackets do – then Phayllos would have chosen a weight of about 1.2 kilograms.[25]

Alberto Minetti and Luca Ardigó from Manchester Metropolitan University looked at the effect of the *halteres* on standing rather than running jumps as this was one theory for the technique suggested at the time. Without any sort of run-up, they found that a pair of *halteres* could increase a three-metre standing jump by around six per cent by increasing the ground reaction force. The higher jumping force increased the take-off speed and the flight distance.[26]

How would the different designs have affected the take-off force? *Halteres* with an overhang at the front are particularly intriguing. Pictures show athletes jumping with the heavy end extended away from them with their arms stretched out. Since the bulk of the mass is just that little bit further away from the axis of rotation in the shoulders, then the moment of inertia would have increased, possibly by as much as seven per cent. This is a really neat trick. The moment of inertia was made larger without making the weights heavier; this would have increased the reaction force at take-off without compromising the acceleration phase of the run-up.

If I was the shopkeeper, I would advise Phayllos to take the heaviest *halteres* he could run with. I would tell him to get ones

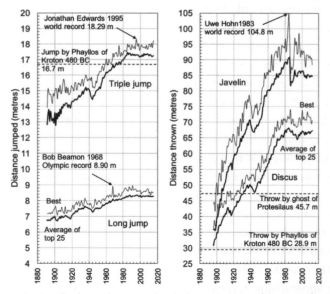

Figure 7. Ancient Greek records compared to the best and average of the top 25 for the triple jump, long jump, discus and javelin. *Data courtesy of Leon Foster.*

with an overhang at the front to increase the effect of his swinging arms and then I would advise him to throw his arms forward during the leap to increase his take-off speed. I might also have asked Phayllos if he would endorse the shop.

The Phayllos conundrum

An epigram written in the second century AD about Phayllos claimed that he jumped 55 ancient Greek feet, or over 16 metres. Given that the area used as a landing pit only went to 50 feet, he would have flown right over it and the story goes that he broke his leg landing on the hard ground beyond.

This puzzled Victorian scholars who were trying to recreate the jump for the 1896 Olympic Games, since it was twice as far as any athlete could jump. The scholars considered everything. Perhaps Phayllos was a truly special athlete who got up to such a high speed that he could jump incredible distances? A quick analysis shows that an athlete would need to run over 65 kilometres per hour to reach the 16-metre distance mark. Even Usain Bolt at his fastest would only be able to jump about ten metres.[27] Running speed wasn't the answer.

Gardiner suggested that a big wooden take-off board might be one part of the solution. He described a British event in 1854 where an athlete used weights to increase his distance from seven to nine metres by jumping off a thick, wide take-off board on the ground. The board would have provided a large target and a firm surface to increase the ground reaction forces, perhaps even acting as a springboard if it flexed a little. Dramatic as this improvement was, it was still well short of Phayllos' record. Multiple standing jumps one after the other were suggested, but this didn't tie in with the idea of a soft 50-foot-long landing pit which would have made this very difficult to do.

I imagine a dusty committee room in a college filled by academics desperate to make a decision on the long jump so it could be included in the approaching Olympic Games. The old saying goes that a camel is a horse designed by committee and it seems that the jump they came up with was the sporting equivalent. They decided that it would have a run-up, two hops and a finally a jump, but no weights.

James Connolly of the USA became the first winner at the Olympic Games in Athens on 4 April 1896 with a distance of 13.71 metres, still about three metres short of Phayllos' world record. The committees for the 1900 and 1904 Olympics created more confusion with the introduction of a standing jump with the feet initially together. It was only in 1908 that the organisers settled on a single hop, a step and a jump and the modern triple jump was born.

A thing for throwing

The ancient Greeks loved throwing things. Ancient Greece was composed of small city states separated by sea or mountains and this natural environment fostered commerce and rivalry in equal measure. Getting an advantage over their adversaries meant that both mental and physical agility was important and the gymnasia became places for learning as well as for physical education. If they weren't trying to sell each other things, they were hurling tree trunks, rocks and spears at each other.

As the technology of war developed, it became clear that throwing things from a distance could soften up the enemy before hand-to-hand combat. Strength and accuracy in the throw became of the utmost importance. Perhaps that's why the word discus, or *diskoi* in ancient Greek, means merely 'a thing for throwing'.

Initially the *diskoi* might have been a rock or any projectile but athletes preferred to use a disc. The idea probably came from the shape of the ingots used to transport metal ore around the Mediterranean. They were often made by scooping a hollow out of sand and pouring in molten metal. This would naturally settle into a pool so that, when solidified, the ingot would be flat on the top and slightly curved underneath. These circular metal ingots evolved into the athlete's thing to throw and might even have been the valuable prize itself.[28]

Unlike jumping weights, discuses were standardised at each event. Olympia had three relatively heavy discuses, all now lost, but around 20 other bronze discuses have been unearthed around Greece. When the committee organised the Olympics in 1896, the specifications they chose were the average from ancient Greece: 21 centimetres in diameter and two kilograms in weight.

The javelin was the other throwing event in the pentathlon and was developed from spears used for fighting. The javelin at ancient Olympia was reputedly the height of a man, the thickness of a finger and made of elder wood. Since Phayllos and his compatriots were around 170 centimetres tall, this made the ancient javelin about a metre shorter than today's equivalent; it would have been about half its mass at around 400 grams.

As a boy, I used to pretend to throw the javelin with bamboo canes from my mother's broad bean plot. They would never go very far as they would wobble, turn end over end or land tip first no more than a few metres away. The reason for this was that the flight was affected by the stick's mass distribution and by natural bumps and roughness on its surface. Too head heavy and my throw would drop short; tail heavy and the front would flip upwards; a slight bend would make it veer off to the side.

You can make a better javelin by machining it straight and choosing the most knot-free wood you can find. But while the ancient Greeks had access to simple lathes and cutting tools, they still had to rely on the natural grain of the wood, which meant there would always be imperfections affecting the flight. Instead, they came up with an ingenious solution which meant that the wood's flaws could be minimised.

They wrapped a short leather thong called an *ankyle* around its middle. Before the throw, the athlete would loop the end of the thong over the forefinger and second finger so that it would act like a slingshot, giving the javelin extra launch speed. As the thong unwound, it naturally made the javelin rotate about its long axis, ensuring that any irregularities in the wood were evened out. Modern gun barrels have a spiral machined down their inside to make bullets rotate and achieve the same effect.

In the early 19th century, Napoleon III heard about the *ankyle* and ordered his generals to carry out tests; they showed that the javelin distance was more than tripled from 20 to 65 metres. More recently, in 2011, Steven Murray and colleagues from Mesa State College in Colorado repeated Napoleon's experiments. They trained 16 young athletes to throw with the ankyle and found that distance increased by about 50 per cent from 20 to 30 metres. Murray didn't comment on whether the javelin actually rotated in flight so this will have to wait for his next experiment.[29]

Sadly, we don't have any distances from ancient Greece to know how well they did with the javelin. However, we do have two records for the ancient discus. The first was by our friend Phayllos who was said to have thrown the discus 95 ancient Greek feet (about 29 metres). The only other recorded throw was by a 15-foot tall ghost of Protesilaus, the first Greek killed in the Trojan War. This friendly apparition apparently threw a discus twice the weight of the one at Olympia just over 46 metres. While we shouldn't put much credence in the throw of a ghost, if it was considered superhuman then it does at least give us an idea of what was considered good.

In the 1896 Olympics, athletes could easily match Phayllos' discus throw but it would take until 1912 for Jim Duncan of the USA to beat Protesilaus the ghost with a massive throw of 47.59 metres. Protesilaus would have been proud – this was designated the first ever IAAF discus world record.

Arte et labore:
the ancient Greek sports industry

Athletics was an important part of life in ancient Greece and every city had its public gymnasiums. They were not quite what we think of today, more like sporting complexes with outdoor spaces for running, jumping and ball sports and with a wrestling school attached. There was often a covered running track for bad weather and a spa with pools. The gymnasium at Delphi had a ten-metre diameter circular pool, two metres deep and with bronze shower heads in the shape of wild animals. Under Roman rule, baths even became heated. The biggest gymnasiums had libraries with schools attached and some countries today still use the word *gymnasium* for their prep schools.

The first place an athlete would go when arriving at the gymnasium would be to the *apodyterion* – the undressing place. This wasn't a changing room as such because there was nothing to change into: as we saw in Chapter 1, Greeks practised their athletics naked. They knew that naked athletics was quirky but they loved the aesthetic of the body and a slathering of oil was just the thing to enhance the golden tan that made them feel like gods.

A whole industry developed around athletics and physical education. Copious amounts of expensive oil were used and there were 'boy rubbers' for hire to administer it for you. Before entering the fray, athletes would enter a special dusting room where exotic powders were applied, each powder having a different purpose. Clay was supposed to cleanse the skin; brick dust enhanced perspiration when the skin was dry; black and yellow earth made the body supple and sleek.

In modern times, the sports industry acts as a barometer for the economy, outpacing growth when times are good and crashing when times are bad. As wealth in the Greek economy grew after Alexander the Great's conquests, it's no surprise that sport prospered. Athletes began to specialise so that wrestlers and boxers became bigger and ex-athletes sought up-and-coming stars to coach. Prize-giving events proliferated and athletes would tour from one to the other, amassing huge fortunes.

By the end of the ancient Olympic Games in AD 392, the Olympics and their sister events were multi-day festivals with running, the pentathlon, bareback horse riding and racing in armour. There was even chariot racing, a bit like having Formula

1 at our modern Olympics. Athletics had become sport and sport was big business. The prizes were valuable: the *stade*, for example, might have had a prize of a 140-litre amphora of olive oil, worth five years' wages of a labourer.

Figure 8. The *hoplitodromos*, or running with armour, was the last foot race to be added to the ancient Olympic Games. The picture shows a recreation at the 2016 Nemean Games. © *Steve Haake.*

The sports industry became really quite sophisticated. There was the equipment to sell: the javelin, the discus, the chariots and the horses, caps to wear during training, slings to carry the discuses in, sponges and soap for washing, strigils for cleaning. The gyms needed undressing room attendants, coaches, trainers, cleaners, boy rubbers, labourers and huge quantities of oil; the allowance for just one athlete going to the gym a few times per week would cost thousands of pounds today. When the Greek and then the Roman empires collapsed, the sports industry died with it. It would take over a thousand years or so for it to appear again.

The puzzle is solved

Phayllos' ancient long jump record of 55 feet was only broken in 1960 when Jozef Schmidt from Poland jumped a massive 55 feet 10¼ inches (17.03 metres), going on to win gold at both the Rome and Tokyo Olympics. With a little irony, in that same year, Harold Harris published a summary of the arguments about Phayllos' jump that had started off the long-jump puzzle in the first place. He pointed out that the original epigram had contained two parts. First came the quote that Phayllos had jumped 50 plus five feet. The second part said that he threw the discus 100 minus five feet. While the jump was exceptional, the discus throw was poor, even by ancient Greek standards. It seems that the neatness of the numbers was what mattered: the use of the nicely rounded numbers five, 50 and 100 made the whole thing suspect. Harris explained that the epigram was a nice ditty with neat, made-up numbers to convey a particular message to the reader, which was: 'You might be good at jumping, but you might be terrible at the discus.' It was the ancient Greeks' way of saying, 'You can't be good at everything.'

After all the analysis, discourse and effort, it seems that the creation of the modern triple jump came from a misunderstanding of the original message. Phayllos probably never jumped 55 feet at all. If they'd known that back in 1896, the triple jump might never have existed.

But the *halteres* showed us the shape of things to come. The way they were designed is just how we design sports equipment today, optimising performance through the correct selection of material, mass, balance point and moment of inertia. The *halteres* were the first personalised sports equipment ever, optimised for the athlete.

It would take the best part of 1,500 years following the fall of the Greek civilisation in the fourth century before anything as organised as the Olympic Games would appear again. The next chapter whisks us there and shows us how our obsession with competition intertwined with technology created the most popular sport in the world.

THREE

Obsession

THE PLAYERS entered to the ecstatic cheers of the fans. Music played loudly and beautiful women entertained the crowd. Some of the men had drunk too much and already looked the worse for wear. Bets were placed on the outcome of the big game. The players stretched their quads and did short knee-jerking sprints to warm up, their mouths dry with fear. A ball was thrown onto the pitch and the noise rose: the most important game of their lives was about to begin.

The year was 918 and the stadium was in Chitchén Itzá, part of an amazing Mayan city in the jungle of the Yucatán peninsula in Mexico. Over 1,000 years ago, alongside enormous stone pyramids, lay a huge ball court measuring 70 by 168 metres. It had vertical walls eight metres high and could easily fit a modern football pitch inside it. The game was played even earlier in 1600 BC by the Olmec, 800 years before the Olympics was even an idea in Zeus' head.

The region was so obsessed with the game that archaeologists have now uncovered more than 1,500 ball courts of different sizes and shapes from Arizona down to Nicaragua in a swathe of Central America known as Mesoamerica. Variants of the game were created during its 3,000 years of existence with different sizes of court and numbers of players. The rules of the game aren't clear, but the players seemed to have used many parts of the body to return the ball including the buttocks, hips, knees, shins, forearms and shoulders. Images show players sliding to their knees to play low balls using their thighs and legs.

Europeans first encountered the game during the Spanish conquest of the Americas in the early 1500s. The conquistadors were used to more sedate ball games back home where the ball was palm-sized, relatively hard and made of cork or stuffed with hair. Columbus quickly shipped some of the balls back to Spain to show off to his compatriots: they were so bouncy, they thought they were alive.

A clue to the secret of the balls can be found in the etymology of the name for the Olmecs. It comes from the combination of two words – *ōlli* and *mēcatl*. *Mēcatl* means 'people'; *ōlli* is the local name for 'rubber'. Olmec simply means 'rubber people'. The game was so popular at its height that Mexico City – Tenochtitlan as it was known then – had thousands of rubber balls made. Sadly, only 50 or so have survived. These have diameters ranging from 13 to 30 centimetres, smaller than a volleyball to bigger than a basketball. The smallest still weighed a hefty 700 grams, the largest a massive seven kilograms (a modern soccer ball weighs 450 grams).

The obsession went deep and even the story describing the beginning of the Mayan civilisation was based on the game.[30] Back in Europe, people were playing in small courtyards with balls that hardly bounced at all but, here, the courts could be huge because the ball was so lively. The Mesoamericans had found out how to make latex rubber robust, stable and elastic, something the rest of the world wouldn't discover until well into the 19th century.

The latex came from the *Castilla elastica* tree. On its own, it wasn't really useable since it cracked, dried out quickly and lacked any bounce. The morning glory vine – *Ipomoea alba* – often entwined itself around the tree. Given time and imagination, someone eventually mixed the two together: this was the secret to vulcanisation. Dorothy Hosler and colleagues from the Massachusetts Institute of Technology recreated the process in 1988 using the following recipe: [31]

Take 750 millilitres of latex and a five-metre length of morning glory vine; strip the leaves and flowers from the vine, coil it, beat it and crush it; squeeze the sap into a vessel to get around 50 millilitres or three tablespoons; mix with the latex and stir for 15 minutes; when almost set, use your hands to form into a ten-centimetre ball.

About twice the diameter of a golf ball, Hosler's ball weighed around 800 grams. When they threw it at the ground, it rebounded two metres into the air. They must have been as delighted as the ancient Mayans.

Figure 9. The ball court at Chitchén Itzá showing one of the rings used for the 'golden goal'. The pitch is 70 by 168 metres. © *Steve Haake*.

The ball game at Chitchén Itzá could last much of the day, but could be won in the Mayan equivalent of a golden goal. High up on each side of the court, still visible today, are two large stone rings poking out of the walls into the playing area like ears. Getting the ball through these would win the game outright and give the player celebrity status. Given that the rings are seven metres off the ground, shots would need both power and accuracy and the ball would have to travel at about 40 kilometres per hour to reach them.

How could they ever score such a goal with so heavy a ball?

The key feature of the rubber ball was its high 'coefficient of restitution', a concept only created by Isaac Newton about 100 years after the Spanish conquest and the fall of Chitchén Itzá. He realised that a body such as a ball never rebounded with the same speed as it landed and formulated a coefficient that was the ratio of the two. If a ball bounced up at the same speed it landed, then the coefficient was one; if the ball didn't bounce at all then the coefficient was zero. Rod Cross, a physicist from the University of Sydney, called balls at these two extremes 'happy' and 'unhappy'. The coefficient of restitution of sports balls falls somewhere between the two with the following values: cricket balls, 0.2; baseballs, 0.6; tennis balls, 0.7; basketballs, 0.8; superballs, 0.9.[32]

The thing about the coefficient of restitution is that it also depends upon what you are bouncing on. The human body isn't particularly good to rebound from, so the Mesoamericans improved the coefficient of restitution by using armour made from stiff leather, wood and even stone (although the main reason was probably to protect from bruising).

The golden goal at Chitchén Itzá was probably scored with the elbow. The faster the ball in, the faster the ball out so it's probable that a teammate set the ball up first by hitting it high into the air as they do in volleyball. The player would then hit the ball first time so that the combination of high speed, a fast swinging arm and high coefficient of restitution would be enough for the ball to reach the rings seven metres off the ground. Players who managed to score a golden goal were celebrated as heroes; some believe the losers lost their heads.[33]

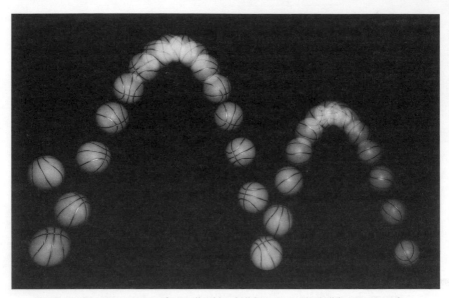

Figure 10. Successive images of a small rubber ball bouncing. The ball loses energy after each successive bounce. The ratio of the speeds before and after impact is the coefficient of restitution; in this case it is about 0.8. © *Michael Maggs, Edit by Richard Bartz.*

Playing in the dark

The Spanish were amazed by the bounciness of the ball but never seemed to enquire how it was made. Perhaps it wasn't right for a good Christian to be associated with a pagan ritual and certainly not one that used a magic ball. Had they discovered how to vulcanise rubber, however, technology might have accelerated throughout Europe, as it would do a couple of centuries later. Certainly, sport would have been very different. Following the demise of the Mesoamerican ball game after the Spanish conquest, it would take another 300 years before the world caught up and rediscovered the secret to vulcanisation.

Following the fall of the Greek and Roman empires, the European world entered the Dark Ages and sport went into the epoch equivalent of half-time. The nearest thing to multi-sport entertainment was the jousting tournament where heavily armed knights on horses would gallop towards each other, lances in hand, scoring points for the level of injury they could inflict. A whole set of rules and regulations appeared such as the *regolamento sopra una giostra* (regulations for a joust), written and distributed in Milan in 1465.[34]

Single points were scored by hitting the shield or upper body; two were gained for hitting the head, three for injuring your opponent so he couldn't continue and four for knocking him off his horse. Every jouster needed to have his lances approved and certified and it seems that knights weren't as chivalrous as we are led to believe. There was a stern warning:

> And the said jousters shall not use any contrivance for cheating either in the saddles or in the weapons, such as hidden girths or other devices which would prevent their being thrown from the saddle.

Cheats would lose their horse and have their reputation ruined by being expelled from the joust in disgrace.

Jousting was really only available to those who could afford a decent horse, armour and attendants; that is, the nobility and the rich. The rest had to find their own pastimes. Monks began to play a ball game called *jeu de paume* in the cloisters of their abbeys which mimicked the two-person rivalry of the joust: one player would serve the ball with their hands from one end of the courtyard to try to get it through an archway defended by his opponent. This game used balls generally stuffed with rags which would make it particularly dull and hard with a coefficient of restitution of only 0.1 or 0.2. Gloves were needed as protection. The square courtyards they played in had covered walkways around their perimeter with archways and the top of the sloped roofs formed part of the playing area.

By the early 16th century, the protective gloves had been replaced by small rackets with rudimentary stringing and a gently looping net was stretched across the middle. Originating in Italy, the game was soon exported to the largest urban population in

Europe – Paris. Of course, the game had to have a name and this was probably derived from the habit of shouting *tenez* before serving the ball.[35] The French were so obsessed with *tenez* that, in Paris, there were said to be more ball courts than churches.

Other ball games began to appear for the masses who only had common land and fields to play on. Because of the hardness of the ball, many involved bats, clubs and rackets. *Paille maille* had a metal reinforced mallet for hitting a wooden ball down a prepared pitch and the fashionable Pall Mall street in London is all that remains of the *paille maille* field from the 17th century. *Soule à la crosse* had a crooked wooden stick to hit a large hard ball in what looked like an early version of hockey and *kolf* was a single stick game using a sheepskin ball filled with wool.

Football throughout this period remained pretty barbaric. The ball was either stuffed with rags and hair or, if you were lucky, had an inflated pig's bladder covered with leather. Most versions of football involved a huge mob of players trying to get the ball from one end of a village to another where injuries were seen as an integral part of the game. Some of these games still exist: in Ashbourne in Derbyshire, locals fight out the game of 'hugball' over the two days of Shrove Tuesday and Ash Wednesday. Teams of hundreds play all day to try to get the ball from the centre of Ashbourne back to their own goal. The goals are three miles apart.

The Industrial Revolution of the 18th century changed the nature of these rural games since they couldn't be played in the busy towns. About a third of people in Britain in 1800 worked on the land, halving to a sixth by the end of the century. The population of Britain had grown to almost 40 million people and people flocked to the towns and cities. London overtook Paris as the world's largest metropolis, growing from one to almost seven million people.

Of course, the attraction of town life was money from new jobs created by factories and mills. The factory owners, however, needed a quick return on their investments and drove their workers hard, often 12 hours a day, six days a week. By the middle of the 19th century, the government had taken notice that green spaces in the rapidly expanding towns were vanishing under streets of brick and coal dust. The General Enclosure Act of 1845 created public parks and gave the population somewhere to play. The Factory Act of 1850 allowed textile workers a half-day holiday on a Saturday, starting at two in the afternoon. Another act in 1867 limited the working week to 60 hours. By the 1880s it seemed that the British had more leisure time than anyone in Europe, local parks in which to take it and, importantly, a bit of spare cash.

The time was ripe for sport to come out of the dark.

Showing off your wares

If there is one event that captured the feeling of the time, it was probably the Great Exhibition of London, held in Hyde Park in 1851. A special crystal palace was designed and built by Joseph Paxton at a cost of £200,000, about £25 million in today's money. Its footprint is still there today, just south of the Serpentine lake. The exhibition had 17,000 exhibitors and 250,000 visitors every month for five months. The aim of the exhibition was for Britain to show off its wares and its superiority; the British Empire had one half of the Crystal Palace, the rest of the world the other.

In the British section, in the North Transept Gallery, under 'Miscellaneous Manufactures and Wares' and surrounded by soap, stuffed birds and fishing tackle, were a dozen exhibits of sports equipment.[36] These displays represented sports technology in the middle of the 19th century, dominated by cricket with names such as Lillywhite & Sons displaying bats, balls, gloves, stumps, shoes and even a catapult-driven bowling machine for training 'in the absence of a first-rate bowler'. William Gilbert's 'Footballs of leather dressed expressly for the purpose' were for rugby rather than association football, which was still 12 years away from being created.

Those of you who've ever been on a sales stand at a modern exhibition will know the fragile atmosphere of companionship and espionage. There were two rival exhibits that were particularly hostile towards each other. They had been obsessively trying to discover what the Mesoamericans had known for three millennia – how to vulcanise rubber. In the British section with the bold title, 'Colonial Possessions, India' was Charles Mackintosh and Company. They had waterproof fabrics, coats, elastic bands and smooth masticated blocks of rubber. The most extravagant piece was a rubber plate with the face of the company's new owner moulded onto it, Thomas Hancock.

In the ostentatiously vast area rented by the United States of America was the Vulcanite Court. The walls were made of rubber, the desk was made of hard rubber, there were smooth rubber curtains, waterproof rubber maps and hydrogen-filled rubber balloons floating gently overhead. And in the middle of the thronging crowd, was the proud inventor himself wearing clothes made of rubber – Charles Goodyear.

The two exhibitors couldn't have been more different: Hancock, the wealthy 65-year-old British cabinet maker turned engineer and businessman; Goodyear, the American shopkeeper turned inventor 15 years his junior and without a shred of business sense.

Rubber shoes had begun to appear from South America during the rubber craze of 1830 when American speculators went crazy to invest in anything remotely rubber. It was soft and pliable, could be shaped and was both waterproof and an insulator. What wasn't to like? Unfortunately, without the secret to vulcanisation, its fatal flaw was soon revealed: too hot and it became soft and sticky with an acrid smell that burnt your nose; too cold and it became brittle and cracked. After the first season of sweltering American temperatures, most rubber products collapsed into a molten heap in the warehouses. So did the market.

For 20 years or more, Hancock had tried unsuccessfully to find the secret of the Mesoamericans. Goodyear had become similarly obsessed and had bet everything on finding the answer – his home, his life and even the health of his family.[37] The secret to vulcanisation was elusive.

Scientific discoveries are often a mix of inspiration and luck. Goodyear's experimental technique was terrible, a seemingly random series of experiments that eventually revealed sulphur as a chemical that would partly stabilise rubber. However, as soon as it got too warm, it would turn soft and tacky again. Goodyear avoided heat at all costs until one day he accidentally left some strips of rubber lying against a stove. When he returned, he found that they were the best samples he'd ever made. While a small amount of heat was raw rubber's downfall, the right amount for the right period of time was vulcanisation's secret.

Rubber on its own consists of long polymer chains which are poorly stabilised in the material and flow over each other. The addition of heat created cross-links of sulphur that joined the longer chains together and stopped them sliding over each other. The long polymer chains could now be stretched until they were straight, with the cross-links bringing the material back to its original length undamaged. Rubber was now a stable, elastic material looking for uses.

In a desperate bid for money, Goodyear sent some of his best samples to Britain to look for funding. Unfortunately, they found their way to Hancock. Goodyear had made a fundamental mistake – he hadn't protected his invention with a patent. Hancock understood the clues – blooms around the edge hinted at sulphur and scorch marks on the surface suggested heat. Hancock applied for a British patent just two weeks before Goodyear was awarded his less valuable US patent. A British court case ensued but, sadly for Goodyear, British and American patent law is quite different. In Britain, all that matters is the patent filing date. In America, what matters is the date of the discovery and the evidence to prove it. It was clear that Hancock had stolen the idea, but all that mattered was that he'd filed the patent first: Goodyear lost.

Goodyear claimed moral ownership of the idea in his 1853 self-published book *Gum-elastic and Its Varieties: With a Detailed Account of its Application and Uses and of the Discovery of Vulcanisation*. This was his way of proclaiming to the world that he had discovered the secret of the vulcanisation process, even if he didn't own the patent. In a rambling and incoherent book, he detailed its many uses, including some that were unlikely to ever see the light of day: bedspreads, dishes, telescopes, books, backgammon boards. Some others were eminently sensible – life-jackets, gas bags and hydrogen balloons.

Perhaps his most sensible suggestion was the rubber football.

The world of sport

As Goodyear was writing his book, mob football was undergoing a conversion in the public schools of Britain. At Eton and Harrow, the kicking and dribbling game was popular, at Rugby and Marlborough the carrying game was more prevalent. All used a ball with an inflated pig's bladder covered in fine-grained strips of leather with a button at each end to hold the stitching together. The natural shape of the bladder meant that the balls were slightly oval in shape and, once assembled, they were inflated by mouth through a clay pipe. Whoever took on the task needed a fair set of lungs and a strong constitution as some of the bladders were still in their smelly green state, fresh from the slaughter.

The small town of Rugby in the middle of England had two football makers of repute. William Gilbert was the older of the two and had exhibited at the Great Exhibition. His younger rival and ex-apprentice, Richard Lindon, had a shop opposite the entrance to Rugby school and made a living making shoes and balls. Lindon cut and stitched the leather, while his wife Rebecca looked after their 17 children and blew up the balls. Sadly, she died from a lung disease, probably contracted from one of the infected pig's bladders.

Lindon searched for a better way to make footballs and one that didn't involve pigs. Whether Lindon had read Goodyear's book or not, he soon hit on the idea of using rubber, not for the whole ball as Goodyear had suggested, but for the bladder. It had two advantages – it was six times stiffer than a natural bladder and 50 times stronger.[38] This meant that the ball could be pumped to a higher pressure to make it livelier in the kick and increase its coefficient of restitution. It was also less prone to bursting when a dozen heavy boys piled onto it. Lindon reinvented the foot pump to blow them up.

He subsequently became ball maker to the universities of Oxford, Cambridge and Dublin and, for the next 50 years, his 'Big Side Match Ball' was the best rugby ball in the business. He never patented his idea so rivals soon copied him, including his neighbour William Gilbert.

The ingredients for modern sport were beginning to assemble. A year after Lindon's invention, in 1863, there was an acrimonious split between the public schools who had been trying to unify the competing codes of football. Eleven clubs met at the Freemasons' Tavern in London in November 1863 to argue over the two main issues. Should they dribble the ball, or should they pick it up and run with it? Should they be allowed to hack down their opponents?

Months of argument followed without agreement and those who preferred the kicking version left to set up the Football Association, copying the 1858 rules of the Sheffield Football Association. The schools who preferred a handling game continued fighting both on and off the pitch and eventually formed the Rugby Football Union in 1871 and the Northern Rugby Football Union in 1895 (the predecessor of the Rugby League).

Figure 11. Richard Lindon holding early footballs c.1880. He invented the rubber bladder and revolutionised football.

The new rubber bladders meant that footballs could be made in large numbers to cater for the soaring demand. It also meant that they could be made round and of any diameter since they were no longer reliant on the physiology of a pig. Lindon enhanced his reputation with the creation of a buttonless ball called the 'Punt-about' which could roll in all directions and was the precursor to the round soccer ball.[39]

Wages in the population grew by 50 per cent in the latter half of the 19th century and workers now had a free afternoon to spend it. What better way to pass an afternoon than go straight from work to the pub for a pie and a pint and then on to a football match that started at three o'clock? In Britain, 'soccer'[40] became the sport of the masses, eclipsing all other versions of football. The schoolboys from the public schools, on the other hand, grew up to become the British Empire's army officers, priests and civil servants and never forgot their sporting roots, exporting their own versions of football to countries that were also ready for mass sport. The colony of Victoria in Australia embraced the idea and gave it a unique Australian flavour with a huge pitch and high-tempo running. In America, Ivy League colleges took a look at rugby and created their own version of American football.

The time was right for indoor sports too. In 1891, Dr James Naismith devised a sport for students bored with indoor gymnastics during the winter. He attached peach baskets to the gallery around the gym and created a team sport using a football: he called it basketball. Clara Baer, a PE teacher from New Orleans, heard of the game and wrote to Naismith asking about it, wanting to create something for women. He sent a sketch showing the outline of the court with dotted lines to show the rough positions of the players. Baer misinterpreted these as dedicated zones that players couldn't leave and created the concept of goal keepers and shooters, wing defence, attack and centres. It seemed to work and the game became known as netball. Other sports were invented too: volleyball for those who found basketball 'too strenuous';[41] water polo for those wanting to play rugby in water; and the Dutch-inspired game of korfball that scandalised the press by encouraging mixed-sex teams and bare ankles. After hundreds of years of informal pastimes, sport became organised and blossomed within a few short decades into the games we see today. Richard Lindon's lack of a patent meant that anybody could copy him and make an inflated ball of any size and any shape. Sports prospered and in the last 50 years of the 19th century, ball sports spread around the world via the British Empire.

I once went to a Dunlop-Slazenger tennis ball factory. At one end of the building were oozing billets of latex, clanking masticators and a distinct smell of pig. (Urea from pigs' urine was used to promote the vulcanisation – smell a tennis ball, particularly when it's wet, and you get a distinct smell of our little piggy friends.) At the other end of the factory, lines of women took bald rubber balls out of pressurised moulds and stuck on dog-bone-shaped felt pieces to give the classic tennis ball cover. While the product was a tennis ball, the place was merely a rubber vulcanising factory. It was very clear to me that without Goodyear's rubber, none of our ball sports would exist the way they do today.

Manipulating the ball

Each country's version of football had its idiosyncrasies, reflected by the ball. But which came first? Did the ball dictate the style of play or did the style of play dictate the ball?

I have to apologise for the use of imperial measurements for the next part since balls were made to the measurements of the time: feet and inches, pounds and ounces. The original ball sold by Lindon and Gilbert in Rugby was 12 inches long, nine inches wide and weighed 12½ ounces. Rugby evolved into a sport where the offside rule meant that only a backwards pass was allowed. This needed a ball that could be grasped with both hands and its width narrowed down to the seven-and-a-half-inch width we have today.

American and Australian football, on the other hand, allowed long distance forward passes which are only really possible with one hand. This promoted an even narrower ball and American and Australian footballs reduced to just under seven inches in width, matching the typical span of a man's hand. Fast runners in Australian football could easily outrun a defender and so a rule was introduced to slow them down which mandated a bounce every 15 metres. A pointy ball was too unpredictable to control so it lost its points – Australian footballs are now slightly rounded and an inch shorter at each end.

To stop arguments about whether a ball was pumped up or not, specifications were set for the internal pressure. Australian footballs have an allowed range of pressures between 9 and 11 pounds per square inch. The range for rugby balls is 9.5 to 11 pounds per square inch while American footballs have a much higher pressure of between 12.5 and 13.5 pounds per square inch. This would make them quite firm to grip. While the internal pressure of the balls might seem a boring technical point, it caused one of the biggest sporting controversies of recent years. The American press called it 'Deflategate' and at its centre was the quarterback for the New England Patriots, Tom Brady.

The date was 18 January 2015. The players entered to the ecstatic cheers of the fans. Music played loudly and beautiful women entertained the crowd. Some of the men had drunk too much and already looked the worse for wear; bets were placed on the outcome of the big game. The players stretched their quads and did short knee-jerking sprints to warm up, their mouths dry with fear. A ball was thrown onto the pitch and the noise rose: the most important game of their lives was about to begin. Human behaviour hasn't really changed all that much over the last 1,000 years.

The temperature might have been down at nine degrees Celsius in the Gillette Stadium that night, but the crowd were feverish with excitement. The New England Patriots and the Indianapolis Colts were about to fight it out for a place in the Super Bowl final. In the match officials' locker room, the referee checked the inflation of the balls, a dozen for each side; each team would use their own

Figure 12. Tom Brady, quarterback of the New England Patriots. © *Keith Allen*.

balls when they were in the offense. Tom Brady requested that his team's balls were inflated as low as possible, close to the minimum of 12.5 pounds per square inch. The Colts preferred something in the middle, at about 13 pounds per square inch.

After they'd been inflated, and against pre-match protocol, the Patriots' ball handler Jim McNally took the two kitbags of footballs from the officials' room to the field, stopping off at the bathroom on the way. The game got under way and by half-time the Patriots were comfortably ahead with a ten-point lead. But the Colts smelled a rat. Their defense intercepted one of Tom Brady's passes and kept hold of the ball. They checked the ball's pressure, suspecting it was below the legal limit and found that it was. At half-time, they requested that the referees check all the Patriots' footballs.

The match officials did what they could in the short time available. They found that the 11 remaining Patriots' balls were all below pressure at around 11.3 pounds per square inch and reinflated them. They measured four Colts balls before they ran out of time, and found they were all just in range. Play resumed but the Patriots improved further and romped home to a 47–7 win.

The scandal broke the next day with accusations that the Patriots had cheated. The NFL started an inquiry and players, coaches and officials were investigated with mobile phones taken for forensic analysis. The evidence was clear: Tom Brady had clearly stated his preference for balls with a lower pressure and all were below the legal limit.

So, who let the air out?

Before we answer that, perhaps we should ask what Tom Brady's reasons for wanting low pressure might have been. What would I have wanted if I were him?

If we go back to the rule specifications, there is surprisingly quite a lot of room for manoeuvre. They allow a diameter around its middle between 6.6 and 6.76 inches, a mass between 14 and 15 ounces and, as we already know, a pressure between 12.5 and

13.5 pounds per square inch. These ranges don't sound like much, but they could make the difference between a successful and unsuccessful pass.

What would I choose to optimise performance?

Firstly, I'd choose the ball with the smallest diameter since it would have five per cent lower air resistance, allowing it to fly further. I'd then choose the heaviest ball. This might be counter-intuitive, but once it was in the air, its inertia would mean that drag would have less effect in slowing it down. Again, it would fly further.

Lastly, I would have the balls at the lowest possible pressure so that my grip would sink into its surface and give me good control during the throw. One way to do this, of course, is to stop off at the bathroom to let some air out. The other simpler way is to let nature take its course: physics dictates that reducing the temperature of a fixed volume of air in the ball also reduces its pressure. A quick calculation using the 'ideal gas law' shows that a Patriots ball at a pressure of 12.5 pounds per square inch at locker-room temperature would reduce to 11.3 pounds per square inch when cooled down to pitch-side temperature.[42] This is exactly what was measured at half-time. Tom Brady or his team didn't need to do anything, as nature would have reduced the air pressure for them.

The Patriots' footballs did have low pressure at half-time, that's undoubtedly true. It's possible that someone let some air out, but the pressure would have dropped through cooling anyway. Circumstantial evidence, however, pointed towards Jim McNally intentionally deflating the balls during his short stop at the bathroom on the way to the pitch.[43] His claim of innocence wasn't helped by the fact he'd called himself 'the deflator' in texts to others allegedly involved in the conspiracy.

But, here's a question I ask my engineering students when faced with ambiguous results: if the results you have in front of you were used to design a plane, would you get in and fly it? This usually focuses the mind. Personally, I wouldn't have boarded the NFL conspiracy plane. However, the NFL believed in their guilt and the Patriots were punished with a $1 million fine. Brady was suspended for four games.

Sport can be obsessive for those who play it and for those who watch it; football is testament to that. Scientists can be obsessive too and can be as competitive as any group of footballers; we'll see this theme repeat itself in the next chapter in a sport that appeared almost out of thin air not long after football.

FOUR

A game of invention

2 OCTOBER 1977: Raquette d'Or Championship, Aix-en-Provence, France. It was blustery, but the weather conditions boded well for the men's final. Although it was a relatively minor tournament of the tour, the draw had delivered an enticing final match: maverick Ilie Nastase against world number two Guillermo Vilas from Argentina, currently on an amazing 46-match winning streak. What the crowd didn't know was that Nastase had a secret weapon.

The previous week, the Frenchman Georges Goven had beaten Nastase in Paris 6-4, 2-6, 6-4. Goven had only beaten him twice before and had only ever taken six sets off him in eight years of competition. This might have been just one of those delightful shocks you get in tennis, but the truth was that Goven had played with a 'spaghetti racket' – this produced so much spin that the ball was often unreturnable. Nastase vowed he would never play against one again.

In 1977, the International Tennis Federation (ITF) had been warned about the racket and wondered how to control it. The racket had two independent planes of strings that could roll over each other rather than one with interlaced strings which were relatively fixed. The ITF asked the University of Brunswick in Germany to carry out tests and they suggested that the ball made one set of strings hit the second set so that each shot was technically a 'double hit', something not allowed by the rules. This was a bit tenuous but it was against the rules and would be enough for them to put a temporary ban on the racket.

In later research, Dr Simon Goodwill, one of the world's leading sports engineers, showed how the spaghetti racket gave up to 50 per cent more spin than a normal racket.[44] He found that when a player used a standard racket in a forehand shot, the ball hit it at an angle and forced a few of the horizontal main strings downwards; these snapped back into place while the ball was still in contact with the string bed to whip the edge of the ball and give it a little extra topspin as it left. The spaghetti racket dramatically increased this effect by allowing the whole plane of strings to slide over each other, a bit like an elastic carpet slipping over a wooden floor. When the plane of strings sprang back into place, the surface of the ball was given a huge upward slice, mimicking a heavy topspin shot.

In the period up to the ITF ban, Nastase acquired a spaghetti racket of his own and used the last opportunity before the ban took hold to use it in the Raquette d'Or final. He might have said he'd never play *against* one, but he'd never said he wouldn't play *with* one. Nastase easily won the first set 6-1. The second set was closer, but still Nastase won it 7-5. Dramatically, Vilas retired from the game, claiming that the strangely spinning ball hurt his elbow. The crowd was incensed and the tennis media went into a frenzy.

The ITF ban on the racket was enshrined in the rules at the ITF AGM the following June. This was a pivotal moment for tennis as, rather surprisingly, it was the first rule ever to be introduced for the racket. This might seem remarkable but, up until that moment, it hadn't needed one.

The original box set

Tennis had already been played for centuries inside monasteries and castles. The hand-ball game of *jeu de paume* played by monks in 16th-century Paris had spread out to reach most parts of the 'civilised' world. Rather than rely on there being a courtyard of the appropriate shape, special indoor courts were constructed that mimicked their origin. They had a net across the middle, a high ceiling, a viewing gallery down one side and a sloping roof to replicate the original cloisters. Shakespeare suggested the balls were filled with human hair and wrote in *Much Ado About Nothing*: 'The barber's man hath been seen with him [Benedick]; and the old ornament of his cheek hath already stuffed tennis balls.'[45]

King Louis XV of France became so exasperated by shoddy workmanship that he set standards for ball manufacture. The ball was made by wrapping strips of cloth tightly into a sphere and covering the whole thing with felt. It weighed 72 to 78 grams and had a diameter between 62 and 66 millimetres, about the size of your palm.

The early protective gloves had been replaced by tennis rackets made from a two-metre-long strip of ash or chestnut. This was boiled or steamed and bent back on itself to give a loop for the head with the long ends bound and glued together to make the handle. A wedge was inserted in the throat to fill in the gap and sheep's gut was used for stringing (cat gut was *never* used).[46]

The game was played off the side and back walls just as in squash and a key tactic was to drop the ball into the corners where it was difficult to scoop out. This prompted two racket design evolutions. Firstly, the head was made squared-off rather than rounded so that it could fit into a corner; secondly, the head was tilted slightly so that when the handle was angled downwards as during a shot, the head sat square on the floor.

I was lucky enough to play this version of tennis in one of Henry VIII's old palaces, Hampton Court, in the west of London. The word 'play' is too grand a word for what actually happened as I think I only hit the ball properly once, maybe twice. The professional was polite and hit the ball gently towards me across the drooping net. I swung the small-headed racket and missed the ball each time as it flashed past underneath my racket. I was so used to the high bounce of a rubber-cored tennis ball that I couldn't overcome my natural swing: the coefficient of restitution of the old-style tennis ball (now filled with cork) was so low that it hardly seemed to leave the floor at all. Eventually, I managed a return and the ball thundered back at high speed, straight into the net.

Just as Queen Victoria came to power in 1830, the future of modern tennis was secured with an unlikely invention – the lawnmower. Edwin Beard Budding had seen bench-mounted machines that were used to cut off stray strands of wool from woollen cloth and realised that something similar could be used to cut grass. His patent had a cylinder roller that drove cutters ahead of it but was so heavy it needed someone to pull as well as push. It became a standard gardening tool for the upwardly mobile Victorian middle classes.

While they initially wanted beautiful lawns to look at, they also wanted to hold garden parties and play games, particularly those that gave men and women a reason to fraternise. The lawns were perfect for croquet, a game probably imported from Ireland. All that was needed was a flat area of grass, metal hoops and mallets to hit the ball with. The game was an instant success and the first all-comers' competition took place not far from Budding's factory in Moreton-in-Marsh in Gloucestershire in 1868.

A small group in Birmingham tried outdoor tennis on their croquet lawn. If the balls hardly bounced inside on a wooden floor, though, they were simply awful on grass. Hollow rubber balls had a much higher coefficient of restitution of about 0.7 and were imported from Germany: they were the perfect solution for tennis on a lawn. The group seem to have kept the concept of lawn tennis to themselves, but the magnificently titled Major Walter Clopton Wingfield soon hit on the same idea.

By 1874, he had applied for a British patent for a new and improved portable court for playing the ancient game of tennis. He bravely called the game 'Sphairistike', a word cobbled together from the Greek for 'ball' and 'game'; the masses weren't convinced and logically called it 'lawn tennis'. If a sport could sulk, that's what the old indoor game of tennis did. To establish its status as the right and proper game of tennis, it renamed itself 'real tennis'.

Wingfield's game was an immediate success with the aristocracy and he sold thousands of box sets at five guineas apiece – about £550 today. His box included four rackets, a bag of rubber balls, nets, lines and a rule booklet. He sold well over £500,000 worth (in today's money) of tennis sets in the first year alone.

Other manufacturers got in on the act and made their own box sets with variants to the equipment and rules so that they wouldn't infringe Wingfield's patent. This caused confusion and tennis went through the same birthing pains that football had recently endured. The Marylebone Cricket Club (the MCC) was one of the only sports bodies around and helped to sort it all out and come up with a unified set of rules. By April 1877, the All England Croquet Club in Wimbledon had added 'Lawn Tennis' to its title and promptly set up a men's tournament. The powers-that-be took control of the game and Wingfield walked away from tennis forever. Instead, he set up a formation cycling team and studied the science of cooking: that's another story.

The creation of a serious championship with prize money of 25 guineas (about £2,700 today) meant that a clear set of rules were needed for anyone taking part. One of the rule makers, J.M. Heathcote, had already suggested that the ball ought to be covered in white flannel to make it easier to control, bounce better and easier to see.[47] The size and shape of the court was fixed at 78 feet long and 27 feet wide with the net at three and a half feet high. The ball was to have a diameter of two and a quarter inches (57 millimetres) and a weight of one and a half ounces (43 grams). Players could use any racket they wanted.

The game prospered, bolstered by private courts and tennis clubs. A competition was suggested between the British Isles and America and the first Davis Cup took place in Massachusetts in 1900. The British were confident of a victory; not only had they invented the game but they also had the best players in the world, or so they thought. The Americans were desperate to beat the old country and set the tone for all subsequent Davis Cups by ruthlessly exploiting home advantage.

The British left home to a fanfare but had no idea of what they were walking into. The British players were renowned for their power game and the American strategy was to neutralise it. The first thing they did was to schedule the matches for the heat of the day when the British players would get tired more easily. Nothing could be done about the rackets since they were individual to the players, but they could do something about the ball. They chose a soft one with a low coefficient of restitution that didn't fly off the racket or the grass particularly well. Then, when they prepared the courts, they left the grass a little longer than the Brits were used to: this slowed the ball down further. One of the British players, Roper Barrett, complained that the grass was twice as long as at home and said that the balls were 'soft and mothery and when served with the American twist came at you like an animated egg-plum'.[48]

The British were trounced 3-0.

Ball standards were introduced so this couldn't happen again. There was, however, still the issue of the courts that could give teams a home advantage. I joined the ITF Technical Commission nine decades later in the 1990s and the issue of court speed in the Davis Cup was still a matter of concern even then. The gold-standard test used an air cannon to fire a ball at 108 kilometres per hour at 16 degrees to the court surface. The speed and direction of the ball was measured by two sets of light gates, one just before and one just after the bounce. The horizontal and vertical speeds were used to determine a 'pace rating' for the court: a slow court had a rating less than 29 while a fast one had a rating more than 45.

The system worked well but it needed an experienced operator, a compressor for the air cannon and three heavy flight cases to carry it around the world. It also cost around £40,000 and there were only a few in existence. Naively, I suggested at a Technical Commission meeting that my research team could easily replace it with a simple portable device. I suggested confidently that it could be finished within a year and would cost just a few hundred pounds. How wrong I was.

The design brief was such that it had to fit into a single flight case under the weight limit of modern low-cost airlines. It had to be simple to use so that an umpire without any technical expertise could use it to check the courts during the pressure of tournament preparation. And finally, it had to agree with the original gold-standard device. We realised that there was a reason it had cost so much – it was difficult to make. Two of my sports engineers, Ben Heller and Terry Senior, set to work to come up with the low-cost alternative. Terry designed and developed a novel miniature air cannon that could be primed with a simple foot pump rather than a compressor. Ben miniaturised the light gates and created hardware to do the analysis that could run without mains power with only a small rechargeable battery.

It took over five years and numerous iterations until the final device was ready for production. The price ended up at about £4,000, so not quite the hundreds of pounds I'd suggested. Our device was called the 'Sprite' and is now used before tournaments to check the courts. I tested a grass surface to mimic the 1900 Davis Cup tie with grass twice as long as at home; it had a pace rating of 27, well into the slow category but still within the rules. I guess that first American win still stands.

Big head

The Nastase–Vilas controversy of 1977 forced the ITF to create a rule for the racket. The ITF Technical Commission looks after these rules today and spends a lot of time poring over research data. Writing a rule can be very tricky: a careless word or phrase can be completely misinterpreted and, before you know it, there's the threat of court action.

In the late 1970s, those making the rules did what they could. They brought in a racket rule that tried to say what they didn't want – spaghetti strings – and what they did want – 'normal' rackets. The problem was that describing 'normal' wasn't easy. The first racket rule was very simple:

The racket shall consist of a frame and stringing. The frame may be of any material, weight, size or shape.

Because of the spaghetti racket affair, the rules on the strings were more detailed:

The strings must be alternately interlaced or bonded where they cross and each string must be connected to the frame.

This meant that the spaghetti racket with its two planes of strings rolling over each other was banned. To make it absolutely clear, a note pointed out in a polite tennis-like way that 'the spirit of the rule was to prevent undue spin on the ball that would result in a change in character of the game'.

As far as the spaghetti racket was concerned, the rule did the trick, and the racket became another exhibit in the Wimbledon Tennis Museum. But a revolution was just around the corner and the ITF would be forced to rewrite the rules again within the next few years.

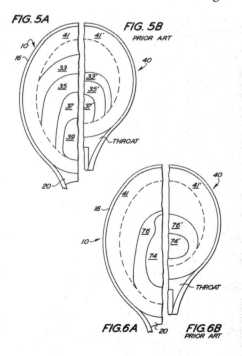

Figure 13. Patent by Howard Head in 1976 for an oversized tennis racket. Bringing the throat down into the handle increased the size of the 'sweetspot'. The wooden racket of American champion Althea Gibson (opposite) is compared to Head's prototype oversize racket.

The instigator of this revolution was Howard Head. He'd had a successful ski company which he named after himself (Head, not Howard). He revolutionised skiing with new skis made from fibreglass and metal, then sold the company to create a new one focusing on tennis. He called it Prince and, in 1976, was granted a patent for a new racket that would change tennis forever. A good designer has the knack of finding solutions to problems that are both simple and elegant: Head's idea for the new Prince racket was both. All he wanted to do was to make the racket head bigger.

This wasn't necessarily a new idea – Frank Donisthorpe had played with an oversized wooden racket at the 1921 and 1922 Wimbledon Championships.[49] The problem with those rackets, however, was that they would buckle if the string tension was too high. The only way to strengthen a wooden racket back then was to make the frame thicker, but this made it too heavy to play with.

Howard Head changed the material to aluminium which was stiffer than wood. He kept the racket the same length and weight, but made the head wider and longer by bringing the lower part of the frame down into the throat. He tested his new racket by firing balls at it using an air cannon at around 160 kilometres per hour and compared the results to those of a standard wooden racket.

When a ball hits a racket, it does two things: firstly, it pushes – translates – the racket backwards; secondly, it makes it rotate about the centre of mass located somewhere near its middle. There is a point on which you can hit the racket where the translation and rotation add together so that the handle doesn't move and the hand feels very little impulse. This is called the centre of percussion.[50] Head explained that with a wooden racket, the centre of percussion was in the throat so you could never actually hit it. By bringing the frame down into the throat *past* the centre of percussion, he made the geometrical centre of the head coincide with it. Players would now hit the centre of percussion which would give less jarring, produce fewer injuries and have a better feel. Howard Head had effectively patented the racket's 'sweet spot'.

The increase in the frame width had an additional consequence that was good for the game. Moving the frame away from the long axis of the handle increased its moment of inertia without increasing its weight (the concept used so successfully by the ancient Greeks for their jumping weights). Head made the racket face ten per cent wider and increased the moment of inertia by 21 per cent. This meant that the racket would twist less in the grip when hitting an off-centre shot and increased the chance that the ball would go back over the net. Learning the game became easier and, within three years, the Prince Classic was the most popular racket on the market.

Other manufacturers didn't take this lying down and invented their own larger-headed rackets just below the size specified by the patent. Dunlop's response to Prince's aluminium racket was to make an injection moulded racket using nylon reinforced with short graphite elements. Injection moulding a racket was difficult to do and the techniques created by Bob Haines and his team at Dunlop were so advanced that they were given an array of Design Council awards and then the Queen's Award for Technical Achievement. Dunlop's mid-size racket, the 200G, was the first graphite racket on the market. It was taken up by John McEnroe in 1983, who used it to overcome a damaged shoulder and go on to win most of the tournaments he entered.[51]

It also signified the beginning of a new revolution. The artisan's skills were no longer needed: from now on, it would be engineers and technicians making the rackets.

The black stuff

At the centenary Wimbledon Championships in 1977, almost all rackets were made of wood, as they had been for the previous 200 years. By 1981, more than two thirds were made from injection moulded plastic or metal.[52] Within five years, every factory making wooden rackets was shut down.

The fear for the ITF was that rackets could now be made longer without making them heavier which might increase serve speeds so much that the nature of the game would change irreversibly. In 1981, they introduced a rule to limit the racket length to 32 inches (81 centimetres).

Tennis racket manufacturers went crazy with invention. For the quarter-century between 1950 and 1975 there had been 64 patents relating to tennis rackets; in the next quarter, there were over 1,000. The star of the show was a new wonder material for sport – carbon fibre.

In 1963, engineers at the Royal Aircraft Establishment in Farnborough tried to produce a plastic with the strength of metal by infusing it with fibres of carbon. Carbon-rich materials were heated until they became liquid and were then extruded through a fine mesh to make long hair-like filaments. These filaments were bound together to make fibres about a millimetre thick which were heat-treated under tension to orientate the carbon molecules and increase their stiffness. Once cooled and cleaned, they were arranged in parallel lines of 100 or more and fed through a bath of epoxy resin which, when cooled, made a cloth of flexible, unidirectional carbon fibres.

By the 1980s, its manufacture had progressed so much that it had begun to appear in Formula 1 cars and aeroplanes. The sheets were layered into moulds which were heated under pressure to make objects of almost any shape or size. Although the sheets were much stiffer in the direction of the fibres than perpendicular to them, they could be orientated to give a wide range of stiffnesses and strengths at any point in the mould. From a racket designer's point of view, this was particularly useful in areas where the stresses might be high, such as at the top of the handle where it joined the head. Equally, if the stresses in a region were low, unnecessary material could be taken away to decrease the weight.

Up to the 1970s, wooden rackets had weighed around 380 grams. Carbon fibre racket designs reduced that by a third.

The tennis establishment's fears that the game would change proved right. The old, heavy, wooden rackets promoted a style where the player stood sideways to the oncoming ball and executed long flowing swings. Matches between masters such as Björn Borg and John McEnroe produced games of invention with volleys, lobs and frequent journeys to the net. The new rackets, however, encouraged fast serves and huge returns from the baseline. The men's game in particular seemed a dull facsimile of what it used to be, with few rallies and games dominated by the server.

Howard Brody, the founder of tennis science, suggested to the ITF that they use the number of tiebreaks as an indication of average serve speeds before the time of speed guns and Hawk-Eye. The theory was that the fastest servers always won their serve so if they played another fast server, a tiebreak was inevitable. The faster the serves, the more likely a tiebreak. The joke was that you might as well miss the first 12 games, go off for your cream tea and come back at 6-6 to watch the real entertainment of the tiebreak.

Analysis was done on data from the Grand Slam tournaments at Wimbledon, the US Open, the Australian Open and the French Open. Tiebreaks – and therefore serve speed – were definitely on the rise and something had to be done. The ITF's first approach was to limit the racket lengths to 29 inches (74 centimetres). The idea was that a longer racket allowed the sweet spot to be further up the racket where it travelled faster when swung. Limiting the length would bring the impact point back down towards the handle and limit serve speeds.

The second thing they did was to introduce a bigger ball. This was just at the time I started to work with them and they asked me to work out the effect of the bigger ball on play. They wanted to keep the weight the same, but to increase its diameter by six per cent to give it more drag to slow it down. I recruited three PhD students to study ball aerodynamics, racket impacts and court surfaces. Prototype larger balls were made by Penn and fired at tennis rackets to see how they rebounded, just as Howard Head had done back in 1976. But this time we were interested in the ball, not the racket. Mathematical models began to be constructed: one for the racket impact, one for the flight and one for the bounce.

One night before a Technical Commission meeting, I met the then technical manager of the ITF, Andrew Coe, in the bar for a catch-up. By the second drink we'd realised that the mathematical models the students had created could be really powerful if we could use them to predict a tennis shot. I stayed up into the night, coupling together their different spreadsheets. The next morning, I had the first version of a prediction tool I called Tennis GUT. This stood for Tennis Grand Unified Theory – my little physics joke and a nod to the unifying physics models of the universe. The idea was that Tennis GUT would allow us to simulate any tennis shot by changing any input parameter: the weight or size of the racket; the coefficient of restitution of the ball; the coefficient of friction of the surface; even the weather.

Most importantly, we could now use Tennis GUT to quantify the difference that the larger ball might make to the game, something that was very difficult to measure experimentally. This might give the ITF a solution to the big serves they felt were blighting the game. Our results showed that the bigger ball would slow the ball enough to give the receiver about two and a half per cent extra time to react to a fast serve. This might not sound much, but it was considered just enough to tip the balance of power away from the big servers.

A new rule for the bigger ball was introduced by the ITF in 2002. Unfortunately, it was never used in a tournament because of the change to a single word during the drafting of the rules. I remember the meeting: it was divided on whether it should be implemented. The draft said that the larger ball *must* be used on fast surfaces such as grass. Those against the idea changed it to say that a larger ball *may* be used on faster surfaces. Somehow, this was agreed and the outcome was that tournaments could choose whether to use the bigger ball or not. They chose not.

Predicting the future

If you look at the rules of tennis today, the larger ball is still in there just in case it's ever needed. One good thing that came out of the research was the creation of Tennis GUT which allowed the ITF to keep one step ahead of the equipment manufacturers. Simon Goodwill turned Tennis GUT into an amazing piece of software and helped the ITF create the test facilities needed to measure the ball and racket parameters needed for the model. The ITF's research labs now have a wind tunnel, ball impact testing, spin testing and even a robot racket power machine that gives them the best tennis testing facilities in the world.

In 2007, we were given access to the Wimbledon museum's store of vintage rackets going all the way back to Major Wingfield's creations of the 1870s. Since we weren't allowed to play with these museum pieces, we put them through Tennis GUT to see how their evolution had affected the game. We found that a 1970s wooden racket could launch a first serve at 225 kilometres per hour; a carbon fibre racket could launch it four per cent faster at 234 kilometres per hour. The receiver had about eight per cent less time to react to the ball from the carbon fibre racket than from the wooden racket.[53] The bigger ball that had been developed to counter this had been predicted to give only about two per cent more reaction time: it wouldn't have been nearly enough to keep the game the same.

But then something remarkable happened: the players with their new rackets began to adapt. André Agassi was perhaps the best example – he seemed to have lightning reactions and used an aggressive open stance style. This contrasted with the old wooden racket style of play where you carefully stepped into the shot and swung through. The lighter carbon fibre racket allowed Agassi to face the oncoming ball and leap from left to right to swat the ball back, using his strong trunk muscles to rotate his body and give the power needed to return the ball.

Grunting caused by the muscular effort of this new technique became a common talking point for commentators, particularly when the women were playing. The rackets were both the problem and the solution. The more forgiving larger-headed rackets meant that off-centre hits still made their way back across the net and the lighter rackets allowed players to get the racket in the right place at the right time. The larger-headed carbon fibre rackets are now the new normal.

Tennis is a great example of how invention can both create a sport and also change its direction. In order to cope with so much change, the International Tennis Federation has had to develop one of the world's best research and testing laboratories. I've been in the lucky position of helping to create some of the systems to bounce, smash and test tennis equipment. And then, when I sit back, I realise that there is a technology without which we couldn't do any of those experiments. It allows us to see what happens when our larger tennis ball flies through the air, when our player runs around the court or when Ilie Nastase swung his spaghetti racket in the Raquette d'Or.

Surprisingly, the development of this technology had sport in mind right from its outset over a century and a half ago. In the next chapter, we'll take a slight diversion and hear the story of a technology that changed how we view sport, one that's so ubiquitous it's almost invisible.

FIVE

Seeing is believing

MY PhD supervisor was Dr Alastair Cochran. He was the technical director of the Royal and Ancient Golf Club of St Andrews – the R&A – and had written the seminal textbook on the science of golf way back in 1968. Called *The Search for the Perfect Swing*, it explained the dynamics of golf clubs, the aerodynamics of balls and almost every aspect of golfing science.[54] What was missing, however, was an understanding of the golf ball's bounce on the green. Alastair had persuaded the R&A to fund a PhD at the Sports Turf Research Institute in Yorkshire. For a project involving physics, the Institute wasn't particularly well resourced, but for a PhD about grass it was perfect: sitting at the top of a hill in a beautiful wooded estate high above Bingley, its main asset was acres of beautiful green hillside.

Alastair wanted me to find out what happened when a ball landed on a green. He'd seen plenty of shots where the ball came almost to a dead halt and other seemingly identical ones where they would screw back as if being pulled by an elastic cord. Was it the grass? Was it the shot?

I was so delighted to be offered the PhD position that it never occurred to me to ask why no one had done it before. But as I sat down in my first months, I couldn't work out how I was going to collect any data. What was I going to do, sit by a green with a notebook for the next three years and watch golf? Perhaps I could get a really good golfer who was so consistent that I could stand on the green and he or she could hit shots to me? But what exactly would I measure? One suggestion was that I could dig up the turf and take it in boxes into the lab to fire balls directly at it with a machine yet to be invented and record it with a video camera I didn't have. But the environment wouldn't be right – I'd have to mock up my own indoor weather system too. What I really needed was a laboratory out on the green, a field laboratory.

As I now know, you're never the first person with a problem and someone has usually been there before you. The trick is to make sure you understand the question and then find the person who's already answered it. Alastair reckoned he knew of someone and suggested I visit him. His name was Harold Edgerton and he was at the Massachusetts Institute of Technology in Boston.

The Doc

Dr Harold Eugene Edgerton, aka 'Doc' Edgerton, sounded like a Wild West hero to me. I was wrong: he was from the East. At the time, I didn't really know who it was I was visiting until someone pointed out that this was the guy who'd taken 'that picture of a milk drop'. You've probably seen the one: it's a startling image of a crown of milk on a red background. The photograph conveys the dynamics of the drop which has come down from high and been caught rebounding from the surface. Edgerton was more than an expert; he was someone who'd worked with the US Army on nuclear explosions and on underwater photography with Jacques Cousteau. No doubt about it, this man was a legend.

In the spring of 1986, I caught a plane to Boston and stayed in a bed and breakfast. My main memory is that all the houses were wooden: everything – wall, ceiling and floor – was made of wood. It occurs to me now that, since this was before the internet, I have no idea how I found the place or even booked it. I arrived at MIT the following morning and reconnoitred the campus to find his office. I walked past his grey metal door a couple of times so I wouldn't be early and then knocked at exactly the appointed time. He must've shouted 'come in' because I entered. Sitting at the other side of a cluttered academic's desk was a balding scientist with a round, welcoming face. As I remember it, he wore a grey suit, a well-worn tie and was framed by a large metal picture window. He seemed old, but then anything over 30 is old to a 22-year-old PhD student. I realise now he was 83. With an encouraging smile and a quizzical raising of his bushy eyebrows, he asked me to explain myself. I told him what I wanted to do – to measure golf ball impacts on golf greens – and his reply was:

"Make a stroboscope, boy."

"Easy to do."

"That should work."

Figure 14. A serve by Gussie Moran recorded at 50 flashes per second (top) and Harold Edgerton (bottom), the scientist who produced the image. The ball was hit to the right at approximately 90 kilometres per hour. © 2010 MIT. *Courtesy of MIT Museum.*

The stroboscope was Doc Edgerton's baby. It is a device that releases regular, short-duration flashes of light. If you run it at high frequency, it seems to give out continuous light but buzzes like a faulty light bulb. Gradually decrease the frequency, however, and it becomes obvious that the light is actually a series of repeated flashes. Edgerton turned this obscure lab device into an everyday tool and then used it to take amazing photographs, many of them of sport. Tennis, golf, even ice skating were objects of Edgerton's interest.

One of his pictures was of a tennis star from the 1940s called Gussie Moran. She scandalised the 1949 Wimbledon tournament by wearing a skirt so short that her frilly knickers were visible. The press photographers went wild trying to get the best shots of her underwear while the All England Lawn Tennis Club worried what the Royals might think. It was so serious that questions were asked in Parliament.

It must have been a bit of a coup for Edgerton to persuade her to pose for one of his shots. The room was blacked out and a dark curtain draped across the wall behind her. Edgerton and Moran probably practised a couple of shots to get the timing right and the experiment would have gone something like this: three, two, one – open camera shutter and strobe on – serve – camera shutter closed – strobe off. The strobe ran at 50 flashes per second and the shutter would have been open for about two thirds of a second so that around 30 images were superimposed on top of each other.

The resulting image shows what happens qualitatively over time. If there is a scale in the picture, then we can also measure distance and even speed. If there isn't a scale, then one trick is to use something of a known size such as the ball – I've done this many times over the years. A tennis ball is 66 millimetres in diameter and in Moran's picture it moved about seven diameters between flashes of the strobe after it was served – about half a metre in a fiftieth of a second. This equates to about 90 kilometres per hour.

The images Edgerton took froze athletes in positions that were previously invisible to the naked eye. It was like seeing sport for the first time. But there were others who'd preceded Edgerton; he was merely the next person to hold the baton. They'd honed their methods way back in the 1800s and their names were Eadweard Muybridge and Étienne-Jules Marey.

Animal mechanism

Eadweard Muybridge and Étienne-Jules Marey were born within weeks of each other in 1830. Although they had the same initials, E.J.M. for Eadweard James Muybridge and Étienne-Jules Marey, the similarity ended there.[55] Muybridge was a restless adventurer, anxious to travel the world and make a name for himself; Marey was a serious scientist treating the world as one large experiment. Fiery Muybridge had minimal education but learnt quickly, was ambitious and never too humble for self-promotion. Marey was well-educated and steady, more interested in using research and his findings to justify his reputation. They would only meet a couple of times, but these brief meetings would influence the world of sport right up to the present day.

Muybridge was born the son of a merchant in the sleepy market town of Kingston upon Thames and was described as being rather mischievous, always getting into something unusual or 'inventing some new trick'.[56] The lure of the USA most likely came to him through the Great Exhibition of 1851. The Americans had been ridiculed for having one of the largest exhibits with the least in it, but had gained grudging admiration for Samuel Colt's new repeating handgun, Goodyear's Vulcanite Court and a rather too-realistic nude sculpture called *The Greek Slave*.[57] Muybridge left Kingston for America the following year, aged 22.

Étienne-Jules Marey spent his early years in the town of Beaune in Burgundy, France. He had a more privileged existence than Muybridge's and was extremely clever, especially with his hands. He breezed through school and set his sights on becoming a doctor; by the time Muybridge was leaving for America, Marey was at medical school in Paris working towards his final exams.

Marey had always been one for creating puzzles, toys and mechanical devices, and during his hospital internship he invented a portable device called a 'sphygmograph'; this was the world's first wrist-worn heart rate monitor. The heart rate monitors you come across today use electronics but Marey's was purely mechanical and showed his genius for instrument design. The sphygmograph had a wooden base the size of a large postcard which was placed flat on the horizontally outstretched arm of the patient. A spring rested on the vein and gently rocked a lever up and down as the blood pulsed by, scratching a mark on a vertical card blackened with soot. A clockwork mechanism moved the card along a linear rail to create a trace much like those you see on hospital monitors today.

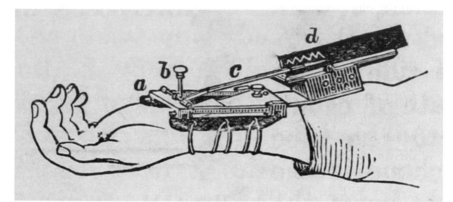

Figure 15. The sphygmograph was the world's first wrist-worn heart rate monitor, designed and built by Étienne-Jules Marey. Secured to the arm by straps, it has an ivory knob which is pressed onto the artery using a steel spring. This causes lever c to bounce up and down, scratching a line on smoked card d carried towards it by a clockwork mechanism. The pulse appears as a trace on the card.

Luckily, it was a commercial success, because Marey had failed his medical exams. The royalties from his invention allowed him independence and Marey set himself up as a researcher in the new field of physiology. Before too long, his attention turned to the analysis of gait, not of humans, but of horses. In the 19th century, horses were still the most important mode of transport and the horse's gait was a regular topic for debate, particularly where racehorses were concerned. The impact of the horse's hooves could be heard as it thundered past, but were too fast to be seen and a common argument was whether the horse's hooves all left the floor at the same time. Was it ridiculous to think of the horse in flight during a gallop?

Marey found horses hard to control when developing his new techniques so went back to humans instead. He designed an ingenious set of shoes to measure foot pressure during running; these had an air chamber in each sole so that they expanded and contracted on each step. The contractions pushed a small amount of air in and out of narrow flexible rubber pipes, which went up the body to a clockwork recording mechanism like the one on his sphygmograph. During a step, air was pushed all the way up the pipes to move a stylus across graph paper leaving a trace representing the pressure under the foot.

It might sound archaic now, but this was state-of-the-art technology and the mechanical equivalent of electronic systems we still use today. Modern pressure insoles now use piezoelectric sheets that fit inside your shoe. The slight bending of the sheet induces an electric charge in the material proportional to the deflection, which is then measured as a change in voltage. The digital traces these make are identical to those taken by Marey.

He measured people walking and running. He showed that both feet left the floor during running and was the first to show that a good performance was related to a short contact time and long flight time. His book *Animal Mechanism*[58] was published in French in 1873 and was one of the first science manuals used to improve performance. Marey returned to the question of a horse's gait.

'There is scarcely any branch of animal mechanics,' he wrote, 'which has given rise to more labour and greater controversy than the question of the paces of the horse.' The reason people couldn't agree on how horses moved, he said, was because they didn't have the right apparatus. How could they argue about whether all four hooves left the ground if they couldn't even measure it? He reckoned he had exactly the right equipment to settle the arguments once and for all.

Figure 16. Marey's ingenious method for measuring gait. Changes in the pressure of an air chamber in the sole of the shoe (bottom right) were transmitted to the stylus of a rotating drum carried by the runner by pipes (left) and recorded on a rotating drum of soot-blackened graph paper. The resulting traces (top right) show the pressures under the feet (C and D) and the motion of the head (O) which came from a third device strapped to the crown.

Marey adapted his human shoes to work on horses and used his graphs to show that a horse's hooves all left the ground at the same time. He employed an artist to work backwards from the graphs to reconstruct what he deduced the positions of the legs and hooves ought to be. What should have been international news was greeted by silence. The problem was that people didn't understand the graphs and the drawings didn't look right: they just weren't convincing.

Marey translated his book into English and moved on.

The flying horse

In 1874, a translation of *Animal Mechanism* found its way across the Atlantic to Leland Stanford, the founder of Stanford University and ex-Governor of California. This is where we find 44-year-old Muybridge. His two decades of travel had been eventful and he'd made a name for himself, but not in the way he might have predicted. He was being tried for murder.

After making the tricky journey across America, he'd settled in San Francisco as a modest publisher of photographs and books. He'd nearly died in a stagecoach crash in which a serious head injury had left him with severe headaches. He returned to London to recuperate, with his doctor advising plenty of walks in the fresh air. Muybridge was a man who was unlikely to waste an opportunity and spent his time learning the science and art of photography as he walked. He returned to San Francisco as a professional photographer with the artistic name *Helios*.

His pictures were dramatic. One had him sitting unconcerned at the edge of a precipice in the Yosemite Valley. His pictures sold well and he found a wife, not an easy task in a gold-rush city where the ratio of men to women was seven to one. Just when everything was going so well, catastrophe struck: he discovered that his wife had been having an affair and the child she'd just given birth to wasn't his own. Muybridge found a gun, hunted down the lover to a remote homestead and shot him dead on the porch. Arrested for murder, he spent four months in jail before the trial in February 1875. The judge directed the jury of 12 married men to give a guilty verdict. They disagreed, saying that, as married men, they would have done the same as Muybridge and that he'd suffered enough. He was acquitted.

During the inevitable divorce proceedings that followed, his wife died suddenly and he put the baby into an orphanage. Muybridge was no longer a youthful man, but grizzled and with snow-white hair, a long, ragged beard and a menacing stare.

Leland Stanford didn't care whether Muybridge was a murderer or not. Marey's book had inspired Stanford to pay for his own research on horses and wanted Muybridge to take pictures that would finally show a horse in flight. Muybridge began to play around with mechanical shutters and decreased the exposure time to only two thousandths of a second – short enough to stop the blurring of a galloping horse. The first photograph they took was of poor quality but had the evidence they were looking for – all the hooves were off the floor. But, as he would do throughout his life, Muybridge would take a shortcut to get what he wanted: he employed an artist to paint a better copy for him to photograph and then pass off as the original. Their critics weren't fooled.

Muybridge and Stanford went back to the drawing board, the latter using his wealth to set up a field laboratory with 12 cameras arranged in a line. Rather than one picture, they would take 12 of them one after the other as the horse galloped by: one of these was bound to capture what they wanted. On 15 June 1878, Stanford's horse, Sallie Gardner, galloped down the track, breaking trip wires across its path to trigger the camera shutters. By the ninth camera, it was spooked by the wires and jumped high into the air. This time, to allay any suggestion of cheating, Muybridge invited the press into the darkroom as he developed the photographic plates in front of their very eyes. Consecutive images of the horse appeared with the ninth picture showing a horse jumping into the air. The press were amazed and, this time, they were completely convinced. Here was the proof that Stanford had been after. The images swept the world and have been going around it ever since.

Muybridge had begun to solve some of the issues that I would encounter during my own research: how to take images out in the field, how to trigger the camera, and how to take measurements from the photographs. The latter was simple: he put numbered vertical lines 21 inches apart on a whitewashed wooden fence on the far side of the track so that when viewed in the photograph, it looked like the horse was running in front of a large ruler.

But people who weren't there that day were still not quite convinced by the still images. The positions of the horse's legs were so outrageous they just couldn't be believed. Muybridge took Marey's suggestion that his own illustrations of horses in *Animal Mechanism* could be put in a popular toy of the time called a zoetrope. This had a ring half a metre or so in diameter with a line of animations on the inside showing consecutive still frames of movement. They had viewing slots between the images so that when the viewer rotated the ring and looked through slots from the outside the brain filled in the gaps between the still images to give the illusion of motion.

Rather than use drawings, however, Muybridge had real photographs. One moment the strange positions of the horse were static and unbelievable. Rotate the zoetrope, however, and the horse would gallop, revealing the truth of his pictures. Ever the showman, Muybridge modified the zoetrope by attaching it to a magic lantern to project the photographs onto a large screen for audiences to see. We take it for granted today that we can take video, pause it whenever we like and replay it in slow motion or at full speed. Muybridge was the first to show how this could be done: his fame was secured.

Human locomotion

Muybridge doubled the number of cameras to 24 and replaced the cumbersome trip wires with an electric timer. This allowed him to move away from horses to people. He took his apparatus to his local gymnasium in San Francisco and took images of athletes running, jumping and throwing. Muybridge embarked on a long lecture tour of America, continuing on to Europe to capitalise on his notoriety. Marey invited him to give a lecture to his scientific colleagues in Paris. Muybridge's long grey beard gave him the look of a distinguished professor and he mesmerised his audience with his animated images of horses, athletes and boxers. Marey's *Animal Mechanism* had been the catalyst for Muybridge to make instantaneous images of horses with Stanford. Now Muybridge became the inspiration for Marey to take his own work to the next level. Muybridge had shown how photography could be used to both understand and measure human motion.

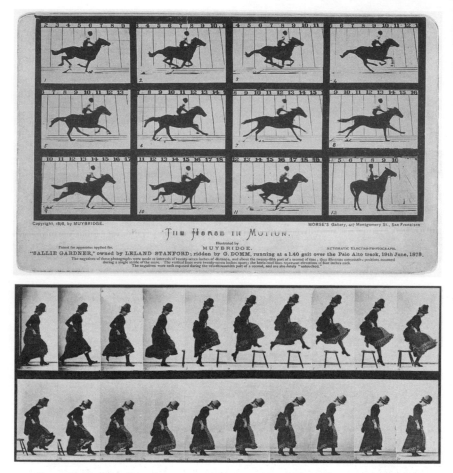

Figure 17. Muybridge's revolutionary motion capture images. Top: 1878 images of Leland Stanford's horse 'Sallie Gardner'. Bottom: images from Muybridge's 1887 book *Animal Locomotion*.

Marey realised that using 12 or 24 cameras in a line was impractical for real experiments. If he was to do any useful analysis, he needed images from a single viewpoint at exact moments in time. By the following year, Marey had developed a single camera system that fitted the bill. It looked just like a rifle with a camera magazine in the hock just above the trigger. This contained a clockwork mechanism that rotated a slotted wheel in front of a photographic plate to give multiple exposures, one on top of the other. To take a photo, he simply took aim and pulled the trigger.

Cleverly, the frame rate of the camera could be increased by increasing the number of slots, while the exposure could be increased by widening them. Marey experimented all summer and printed his images of men running in La Nature later that year; these were superior to Muybridge's in almost every way.

Figure 18. Marey improved Muybridge's technique by using a single camera with a moving film plate coupled to a fast shutter. He created 'chronophotography' by making his subjects wear black clothing with reflective buttons at the joints and white bands between them.

Marey then did something that all biomechanists today would immediately recognise – he made his subjects wear a jet-black suit and marked their joints with bright metal buttons. He then used white bands between them to mark out the limb segments, making his subjects run in shadow so that all that was recorded was a stick figure in motion. Marey dubbed his new technique 'chronophotography'.

The opportunity to analyse humans in various sporting poses wasn't lost either on Marey or Muybridge. Marey went from strength to strength and built a Station Physiologique in the Bois de Boulogne in Paris with a large grant from the government; it now sits beneath the Roland Garros tennis stadium where the French Open is held. Muybridge was commissioned by the University of Pennsylvania to take images of people during physical activity and then travelled the world presenting them. They are so inspirational that, over 100 years on, his book on *Animal Locomotion* is still in print.

Muybridge was eventually overtaken by the movie camera and never seemed to be able to move on from his archaic methods. He died in 1904 in Kingston upon Thames in his back garden, apparently semi-naked while trying to create a scale model of the Great Lakes.

The future is golf

Back in the 1980s, I returned from America with Harold Edgerton's advice burning in my ears. My supervisor had been busy and had persuaded Acushnet, the company that makes Titleist golf balls, to give me an equipment budget in return for first sight of the research results.

My three inspirations – Muybridge, Marey and Edgerton – had shown me that it was possible to set up a field laboratory and measure what I wanted using photography. They had a few key principles: trigger the camera to take the photograph at the right time; use a short exposure time that will freeze your fast-moving subject; and lastly, put something in the picture to give you scale.

To set up my experiment, I first needed to know the speed at which golf balls landed on the green. Acushnet supplied me with a state-of-the-art computer model of a golf ball trajectory that gave surprising results. A shot from a driver landed at about 97 kilometres per hour and with a huge 2,000–3,000 revolutions per minute of backspin (in comparison, a vinyl LP record rotates at a lowly 33⅓ revolutions per minute). A shorter, slower shot, such as a nine-iron, landed at around 65 kilometres per hour but with even greater backspin of 9,000–10,000 revolutions per minute.

The backspin was key to my whole research question – on some greens the spin would make the ball come to a dead halt while, on others, the ball would bounce forwards but then screw back on the second bounce. How was I going to measure any of this?

Muybridge's line of multiple cameras was unrealistic and, while a photographic gun like Marey's would have been wonderful, I didn't quite have the funds to get one made. I decided to go with the Edgerton approach and use a stroboscope. I worked out that I'd have to take at least 200 images per second to get the ball landing and rebounding from the turf on a single picture. I bought the highest-powered strobe I could afford and a beautiful large-format Bronica camera with interchangeable backs, one for Polaroid and the other for roll film. Over half my budget was gone.

Of course, I still had the tricky problem of getting a ball to land on the green in the right place at the right time. Then I heard of a baseball pitching machine which could project balls at any speed and spin. It was a bit of an engineering monster: it had two counter-rotating wheels with a small gap between them so that when a ball rolled down a chute between them it was squeezed out the other side with a pop. The faster the wheels, the faster the ball. I could orientate the machine so that the wheels were vertical and make the lower wheel rotate faster than the top one to give the ball backspin.

I used some of my budget to modify it to bring the wheels in closer together so that it could fire golf balls. I test fired the balls into the air to see what would happen: the balls sailed off into the next field, never to be seen again. I pointed the ball-firing machine directly downwards at the ground. The ball thudded into the grass and bounced forwards a few times, leaving a deep divot in the turf. I poked it with a screwdriver and stamped on it, hoping no one could see the damage I'd done.

I had numerous practical problems to solve. The bowling machine and the stroboscope needed power which was usually a long way from a golf green: I bought a one and a half kilowatt petrol generator. The sun washed out the pictures: I built a black tent to fire the ball into. The deep pitch marks meant that I had to put everything on wheels so that the whole apparatus could be shuffled along by a few centimetres after each shot. I spent a couple of pounds on a divot repair tool, since there's nothing like a long line of pitch marks across the 18th green to annoy a greenkeeper.

Each decision had a knock-on effect and I ended up with so much equipment that I needed a large van to transport my field laboratory around in. I was so excited on my first field trip that I scraped the side of the van down the tall stone gateposts to the car park of the labs. Everyone rushed to the windows to see me desperately trying to reverse the severely dented van back off the post. I feel the embarrassment even now.

Eventually, I found myself on my first golf green. After a hard day's work, I'd taken only 30 pictures of impacts. I didn't dare send them out to be developed professionally in case they were spoiled, so I developed them myself in the Institute's darkroom. I still remember the relief as I held up the strip of negatives in the red light to see beautiful stroboscopic images of golf balls hitting a golf green. I'm still amazed to this day that the experiment actually worked.

Each picture had about 20 overlapping images showing the ball rebounding from the turf, and you could even see little bits of grass and sand flying into the air. I put lines on the golf balls so that I could measure their spin and I copied Muybridge's idea of a scale in the image by double-exposing a grid placed down the plane of the impact, removing it before firing in the ball.

It was a slow process but by the end of the PhD I'd visited 20 or so golf courses and recorded about 1,000 simulated shots from a driver, a five-iron and a nine-iron. My conclusion? That the soft inland greens of Britain forgive poorer golfers. On these greens, even if you're not very good at putting spin on the ball, it will pretty much always come to a stop on the first bounce. A seaside links course, on the other hand, will really sort out the good players from the bad. If you get enough spin on the ball, it will check on the first bounce and rebound forwards, but retain some backspin. On the next bounce, the ball will stop dead and even screw back along the green.

Each green construction has a spin threshold where the ball can retain this backspin after impact. This depends upon the grass type, soil composition, moisture content and maintenance regime. The advice to greenkeepers was to construct the greens with sand, give them good drainage and sow fine, drought-resistant grasses like fescues and bents; annual meadow grass was a scourge to be avoided.

Three years of work in two paragraphs.

Through the camera lens

After my PhD, I ended up in Sheffield in the north of England as a lecturer in mechanical engineering. I didn't have the patience for analysing nuts and bolts and naturally gravitated back to the more exciting subject of sport, setting up what grew to be the largest academic sports engineering research centre in the world (probably).

By the mid-1990s I had my first PhD student, Matt Carré, who used a field lab like mine to look at cricket rather than golf (another sport I can't play). The big advance with Matt's method was the replacement of the film camera with the newly developed digital camera. Kodak used standard cameras such as Nikon and Canon and replaced the film with a light-sensitive computer chip. The first one I bought, at a vast cost of £8,000 (about £14,000 today) had an array of 1,012 by 1,524 picture elements – or pixels for short. This was the best digital camera in the world with a total of 1.5 million pixels. In comparison, my current phone has a superb camera with a 12 megapixel sensor; there is even a camera out there with 50 megapixels.

While I'd had to measure all my images manually using a projector and paper, Matt could use a computer to do all the hard work with the digital photographs using clever image processing algorithms. He tripled the number of impacts I'd studied and gave cricket pitch groundsmen information to help them create the perfect cricket wicket. I moved on to tennis and soccer and, by this time, digital cameras could now store video. The sensors were so sensitive to light that 200 pictures per second were easily possible. The biggest problem now was the length of time it took to download the huge files and the cumbersome hard drives needed to store them.

My PhD might well have been one of the last of the sports-related experiments to use still cameras with real film and I'm a little sad that my PhD students won't experience the fear and delight of developing a film in the searing blackness of a darkroom, surrounded by the pungent smell of chemicals. However, the upside is that they can now create images faster and more accurately than I ever could. Their task now is not to figure out how to take the pictures in the first place, but how to analyse them in a way that exploits the faster and higher-quality digital cameras that continually hit the market. They're not really cameras any more, they're computer chips with a lens on the front.

The analysis of human motion suddenly became easier. My current motion capture laboratory has over 30 digital cameras all linked together to give instantaneous motion capture of people running, jumping and gyrating across our tennis court-sized space. These motion capture systems are the digital equivalent of Marey's; while he used bright buttons to track different parts of the body, we now use small spheres a few millimetres in diameter coated with the reflective tape you see on jackets worn by the emergency services. The cameras record the position of the markers in three dimensions and the computers and algorithms are so slick that automatically generated skeletons recreate on screen the movement of the athletes before your eyes.

Once video cameras became digital, it was inevitable they would appear in professional sport. One of the most well known is Hawk-Eye. Ten cameras mounted around a tennis court allow triangulation from three or more to give the three-dimensional position of the ball. Hawk-Eye's business plan wasn't necessarily to sell to the sports themselves, but to the TV channels looking to liven up their coverage. A bizarre scenario quickly developed where Hawk-Eye showed line-calling decisions on TV which didn't agree with the officials in the stadium. Tennis had the untenable situation where the audience at home had a better view of what was going on than the umpire on the court.

The International Tennis Federation was asked in 2003 to assess whether Hawk-Eye, or indeed any other system, was good enough for official line calling. I soon found myself with Jamie Capel-Davies, (now technical manager of the ITF) on my hands and knees staring at tennis court lines sprinkled with talcum powder. We would fire a ball directly at the line using an air cannon and record what happened using a high-speed camera running at 2,000 pictures per second. We would peer into the talcum powder to see the shape of the marks left by the ball and found that it would hit the ground, flatten and then slide along, leaving an elongated oval in the powder. By eye, a ball would easily miss the line and then slide more than 20 centimetres before rebounding. It looked a long way out, but watching on the slow-motion video showed it had just touched the line. By eye – OUT; by camera – IN. The question was, what would Hawk-Eye give?

Hawk-Eye were always a little coy about their accuracy and their on-court graphics would claim measurements to the millimetre. Was this really possible? A fast first serve hits the court at about 108 kilometres per hour. To measure to one millimetre accuracy would require a camera with a frame rate of 30,000 frames per second. This wasn't the sort of high-end camera that Hawk-Eye were using: their cameras recorded at a mere 60 frames per second. At this frame rate, a first serve blasted down towards the tennis court surface would move about half a metre in the air between picture frames, not a millimetre. The chances of catching a ball as it hit the ground would be remote, so how could Hawk-Eye claim anything near this accuracy?

Their system will have technical tricks to make it work, the sort of things you only realise once you try to do it: what to do when the sun creates shadows; how to calibrate the corner furthest from the cameras; how to adjust for camera wobble when the stands erupt into applause. The actual algorithms are commercial secrets but this is my guess to their approach. Firstly, they would probably do a quick sweep of all the frames from all the cameras to find the fastest-moving objects. This would be the ball, the movement of a player's arm or leg, someone in the crowd, maybe even a pigeon flying past. Only the ball will have anything like a trajectory pattern so it would be relatively easy to identify and this would allow them to ignore the rest. Focusing on the trajectory, they would combine the data from at least three different cameras to get the three-dimensional position of the ball, 60 times every second around half a metre apart to give a join-the-dots curve of the ball's flight.

The next thing they'd do is put a mathematical best-fit line through the dots to give a continuous trajectory. With this they can now project a line down onto the ground to get the location of the impact. The same can be done for the rebound, and the distance between the impact and rebound point on the ground gives the slide during contact. Since they have ten cameras, they can repeat this with multiple combinations of any three cameras to give up to 120 versions of the same trajectory. Averaging these would give an excellent guess of where the ball hit the turf.

Until the ITF testing, I suspect Hawk-Eye didn't know the true accuracy of their system themselves. To be approved for tournament use, Hawk-Eye had to go through a rigorous set of on-court testing which was then followed by shadow testing during a match using real human line judges. This took weeks to complete. Ultimately, the system had to agree every time with the line judges, with an average accuracy of plus or minus five millimetres as measured by ultra-high-speed cameras. If any measurement was more than ten millimetres out, the system failed. This was a pretty tough task and there were inevitably some teething problems but, in October 2005, Hawk-Eye became the first automatic line-calling system in tennis.

Hawk-Eye tracked the ball pretty successfully most of the time but still had technicians on standby to overrule anything that might go awry. It was inevitable that there would be a decision that would cause controversy and it happened during the fourth set of the Wimbledon final in 2007 between Roger Federer and Rafael Nadal. Federer was two sets to one ahead, but two games down in the fourth set and serving at 30-30. It was a crunch moment. If Federer won the point he would probably go on to win the set and the match. If Nadal won the point, he might just save himself and force a fifth set. Federer's first serve landed in but Nadal's return was called long. The BBC cameras and the commentators in the studio agreed it was out. Nadal, however, used one of his Hawk-Eye appeals and dramatically, the system called it *IN* by just a millimetre.

The normally cool Federer lost it: he swore loudly enough to be heard by the crowd and asked the umpire to switch Hawk-Eye off, shouting, "It's just killing me today. How in the world was that ball in?"

Federer lost his rhythm and Nadal went on to win the set, tying the match at two sets each. Federer pulled back his form in the final set and went on to win the match, which was perhaps lucky, otherwise the controversy might still be going on. But the doubt was out there – was Hawk-Eye as good as we thought?

The problem comes down to what we expect Hawk-Eye to do. Any measurement system will always have a certain level of error and Hawk-Eye have gone on record to claim that their accuracy was to within about three millimetres.[59] In engineering, this usually means plus or minus three millimetres so that a ball called *IN* by one millimetre could actually be anything from *four* millimetres *IN* to *two* millimetres *OUT*. All Hawk-Eye can actually do is report that there is a higher *probability* that the ball is *IN* than *OUT*.

Hawk-Eye try to get around this by showing a computer-generated animation of the ball hitting the surface with an elongated pitch mark. They could put a fuzzy edge around the pitch mark to show their mathematical uncertainty as some have suggested, but then there'd be a constant debate about close calls which was precisely the problem that Hawk-Eye was supposed to alleviate.

But now another system has finally been approved by the ITF, called FOXTENN. FOXTENN has taken advantage of the fact that digital cameras have improved since Hawk-Eye started out, and use cameras running at 2,500 frames per second. These look right across the service lines and are supplemented with lasers at ground level. If there is a close call, they use the lasers to work out if the ball was in or out and then show real images of the ball hitting the line for everyone to see: after all, seeing is believing.[60] Even Federer would be satisfied with that.

These camera systems have now been adapted to track players. The Marey approach would be to use markers on the player, but these are impractical in play. Instead, software is written to track the whole player – 'blob tracking' as it is known. The key problem is to figure out where the centre of the player is. For an adult, this is not far off the belly button, but if our player bends or twists, this centre of mass moves around; it can even move outside the body if the player bends over enough.

One of my PhD students, Marcus Dunn, worked with the ITF to track tennis players using a single camera high up in the stand. He managed to track the player but then went one step further and tracked the player's footsteps too. He was quite low-key about what he'd done, but he'd actually invented a low-cost gait analyser that Marey would have been proud of. Now, not only do we have the player's position, but we also have step length, step width and stride frequency. This can be used for any sport with any step pattern. We have a system installed in our local sports stadium and even one in the local hospital, helping doctors analyse abnormal gait patterns.

When Marey proved with his graphs that a horse was temporarily in flight during the gallop, people didn't believe it because they couldn't see it. Muybridge provided the images, but even then the limb contortions were so strange that people didn't really believe him either. It was only when he animated the still pictures into a moving image – the first movie – that people finally believed what they saw. This still applies today. As we'll see later, my research centre works closely with British Olympic teams to improve performance; whatever we measure, it invariably includes video because the coaches and athletes want to see things with their own eyes.

Our next chapter returns to our chronology of sport in which we will use the camera techniques developed by Muybridge, Marey and Edgerton. The sport we'll look at has technology in its genes and, along with tennis, led the sports technology revolution of the late 20th century.

SIX

Taking the rough with the smooth

NOT TOO long after I became an academic, in 1997, I found myself in the foyer of Callaway Golf's new golf ball division in the small town of Carlsbad in southern California. Starting out in textiles, Ely Callaway had retired aged 52 to start up his own Californian vineyard. When he sold his wine company in 1981 he retired again but got bitten by the golf bug. Within three years he'd used the profits from his vineyard to create Callaway Golf which went on to sell one of the world's most successful drivers, the Big Bertha, named after a gun from the First World War. Ely Callaway had always seen profit as a way to invest in his next big thing and the Big Bertha had given him even more to play with: he was now ready for another adventure.

Dave Felker was the head of the new golf ball division and he breezed into the foyer. Shaking my hand, he explained that he had the task of assembling the skills and expertise to create Callaway's first ever golf ball. His objective was to develop a ball that was 'demonstrably superior and pleasingly different', Ely Callaway's catchphrase. Not only that, but he had to deliver it within three years, starting from a blank sheet of paper.

I dug out my PhD thesis and spent ten minutes telling him about my research. We talked about my plans and I mentioned I was visiting another golf ball company over on the East Coast. He came straight out with it: "Don't go to them," he said, "sign up exclusively with Callaway as a consultant." Next moment, I was out of the door with a contract in my hand, and I hadn't even made it past reception. That was pleasingly different, I thought. The contract could be signed once I'd got back to them with a charge-out rate for my time. I consulted a few friends and colleagues and plucked a day-rate out of thin air, one I thought at the time was distinctly outrageous. They accepted without a murmur and I realised I'd probably undersold myself.

Callaway launched their golf balls in 2000, right on time: they were called Rule 35. The ball came in two versions, 'soft' and 'firm' in helpful blue and red boxes. The idea was that the ball conformed to the 34 rules of golf set by the United States Golf Association with Callaway's Rule 35 added on the end. Rule 35 was 'enjoy yourself'. By the end of the year, Callaway had sold $35 million worth of balls.[61]

The part I played in the new Callaway golf ball was minimal, as they weren't really interested in what happened when the ball landed on the green. But the effort that went into designing and developing the ball was immense and mirrored in just a few short years the 100 years of development that had preceded it.

A youngster with pimples

The birthplace of golf is disputed. The Scots claim it, the Dutch claim it and, if you go back far enough, even the ancient Greeks and Romans claim it. The fact is, if you have a small ball, at some point someone will pick up a stick and start whacking it. Golf balls had originally been made from turned boxwood, but by the 1600s were made by stuffing a stitched textile cover with feathers. Four pieces of leather were soaked to make them pliable and stitched loosely inside out, leaving a small slit to turn them back the right way around. The slit was then used to poke in enough boiled goose feathers 'to fill a tall hat' and the resultant ball stitched shut. They became known as 'featheries'.

As the ball dried out, the leather contracted and compressed the drying feathers to create a hard ball. The best ball makers were originally from Holland but then Scottish ball makers around Edinburgh began to learn the trade. They could only make around three to four per day with the best balls costing four or five shillings each; this is about £25 today and similar to the cost of a dozen mid-range modern golf balls. The problem with the feathery was that it became unplayable when it got wet since the leather became soft again. A poor shot with a sharp iron could also split the ball open with a catastrophic puff of feathers. A newly purchased feathery, however, was a fine thing and, if it was oiled and hit properly, it could easily fly 160 metres.

The feathery industry lasted at least two centuries but encountered a revolution in the 1840s. The story goes like this. A clergyman in St Andrews by the name of Dr Paterson received a statue of the Hindu god Vishnu from a missionary in Singapore. It was nestled in black-brown shavings of gutta-percha, a hard gum not too dissimilar to rubber latex and coming from the sapodilla tree of South East Asia. Paterson put the statue on a shelf and wondered what to do with the shavings.

Gutta-percha had been around for some time and could be moulded by simply heating it in hot water to make things like handles, jewellery and furniture. It was waterproof and Paterson decided to use it to cover the soles of the family's boots. He heated it, rolled it out like pastry and glued it in flat sheets to their shoes. When the shoes wore out, only the hard-wearing gutta-percha was left.

Paterson's son Robert was a golf obsessive but without the means to buy the expensive feathery balls. What if he could make his own? The gutta-percha seemed a rather clever solution. He boiled up what remained of his boot soles and rolled them into inch and a half spheres in his mother's kitchen. He painted the ball white just like a feathery and took it out onto the links of St Andrews for a trial. The ball flew well, even if it felt a bit hard, but shattered into pieces on the next shot. At cool temperatures, gutta-percha is brittle and any cracks lead to catastrophic failure. Robert retrieved the shards of gutta-percha and took them back to the kitchen, heated them up again and rolled out the ball more carefully this time. The ball lasted longer but still eventually broke into pieces.

Robert tinkered on and let his brother in on the secret. When Robert finished his studies and emigrated to America to found a Bible college, his brother continued with the experiments. He improved the manufacturing process, removed air bubbles and cured them for weeks. He even stamped 'Paterson's new composite' on them, adding the word 'patent', even though he didn't have one.

The ball began to be used and by 1848, other players had begun to copy Paterson; they were a fifth of the price of featheries and immediately overtook them in popularity. At first, the feathery makers feared for their livelihoods, but then found that demand for 'gutties', as they were now called, drove the market for other things such as golf clubs, coaching and caddies. And the advantage of the gutty was that they were quicker to make and if they were damaged could be put back into boiling water to soften out the nicks.

Then an exciting thing happened. The gutties were generally harsher to play with than the featheries and much more difficult to control, but people put up with them because they were so cheap. Initially they veered off strangely during flight, but people began to notice that they flew much better at the end of the day than the beginning. Players suspected that this was due to the nicks and scratches and began to keep them as they were rather than fix them. Eventually, they began to mark them *intentionally* before play had even started.

Willie Dunn, a ball maker from the small town of Musselburgh, just east of Edinburgh, went one step further and used a metal mould with surface markings to compress the hot gutta-percha into a patterned ball. All sorts of surface designs started to appear, one of the most popular being the bramble pattern where the surface was raised with pimples like a blackberry. The ball flew extremely well.

The ball was cheap, even if it was still a bit brittle and likely to shatter into pieces. A rule was made to take this into account and the players put up with the harsh feel and sound. But, like waiting for buses, another revolution was just around the corner, fuelled by the application of science and technology.

Spin doctor

Some of the scientific greats have had an interest in sport, not least because they could use it to explain the fundamental laws of physics. Back in 1671, Sir Isaac Newton was at Cambridge University working on optics. He wrote to his friend Henry Oldenburg about how tennis balls swerved in flight when given spin and explained how unequal air pressure either side of the ball would make this happen. Lord Rayleigh, a physicist with a Nobel Prize, found time in 1877 to write a paper called 'The Irregular Flight of a Tennis Ball'. One of his students, Joseph J. Thomson, would win the Nobel Prize for discovering the electron and would later write a paper in 1910 on 'The Dynamics of a Golf Ball', comparing its trajectories to those of electrons in a cathode ray tube.

Peter Guthrie Tait was another such scientist. Born in Dalkeith, Scotland in 1831, just a year after Eadweard Muybridge and Étienne-Jules Marey, Tait was educated at the universities of Edinburgh and Cambridge, going on to write seminal texts on the physical sciences. He was so prolific that he wrote a journal paper every 44 days of his academic career. By 1860, he was Professor of Natural Philosophy at the University of Edinburgh, a post he would hold until his death 41 years later. He was also an avid golfer and would visit St Andrews to play as many rounds as he could in one day; his record was five.

Tait took a keen interest in the dynamics of golf and wrote 13 papers on it. His third son Freddie became an accomplished golfer who won the Amateur Championship twice and helped his father with golf experiments in his basement. Very quickly, Tait realised that 'underspin' was a key feature of golf; today, we call this backspin. He explained that it was the underspin that caused a long drive to have a slightly concave upward path, seemingly hanging in the air before then dropping steeply at the end of its flight.

Critics were aghast at Tait's conclusions. How dare he suggest that they put spin on the ball, they said; a proper shot should be hit clean and straight with no spin at all. Tait ignored them and continued to formulate his mathematical models. This must have been a long and tedious task, taking months to do the repeated calculations to plot a single trajectory (that's what assistants are for, I guess). He concluded that the underspin was caused by the ball's interaction with the club and that it produced a lift force that counteracted gravity. The more spin, the more lift. He estimated the ball's launch speed to be about 240 feet per second (over 260 kilometres per hour), remarkably close to the standard now used for golf ball testing by the USGA and R&A.[62]

The experiment I like most is the one in which he measured spin. He took a long flat tape and attached one end to the side of the ball and the other to the floor. He then had Freddie hit the ball into a clay pendulum after which they counted the number of twists in the tape. His spin measurements came out between 1,800 and 3,600 revolutions per minute, similar to the values I was given by Acushnet during my PhD. Tait suggested radically that clubs ought to have grooves on their faces to improve the spin imparted to the ball, a feature we now take for granted.

His experiments with his son must have been fun but, sadly, Freddie died in the Boer War in February 1900. His father's heart was broken, his health deteriorated and he died later the same year, leaving us with volumes of papers on physics and the foundation for our understanding of golf ball dynamics.

The Haskell ball

When I was at school in the 1970s, two of the scarier boys from class found an old golf ball on the school playing fields. They cut it up and found inside a tight coiled-up ball of elastic thread: one snip and 40 metres of elastic sprang out. After school, they stretched the band either side of the main road, holding up the taut elastic band so that it pinged back the aerials of the unsuspecting cars zooming past. They dropped it and ran when a bus came. I was desperate to get my hands on a golf ball and see what was in it. Eventually I did, and I took it to my father's garden shed to chop it up. Right in the middle of the ball, surrounded by elastic thread, I found a small sack of white fluid. Of course, I split it open and only just resisted the temptation to taste it. I'm glad I didn't: early golf ball manufacturers had tried water, treacle, honey, castor oil, glycerine, caustic soda and mercury. I never did find out what was in it.

The way these balls were invented is contained in different versions of the same anecdote involving two characters living in Ohio at the turn of the 20th century, at about the same time Tait was doing his golf ball experiments in Edinburgh. They were called Coburn Haskell and Bertram Work. Haskell was a relatively rich 30-something from Cleveland selling bikes for his father-in-law, while Work ran the Goodrich rubber factory in nearby Akron. The company made tyres, raincoats, rubber blankets and a small number of gutta-percha golf balls.

Haskell and Work were golf partners. Haskell felt he was the worst player in their foursome and thought it would be great to have a ball that gave him a bit of an advantage. There are many versions of the next part which you can read in John Martin's excellent book *The Curious History of the Golf Ball*.[63] The most plausible one is that Haskell visited Work at the Goodrich plant and picked up some stray elastic thread. He began to wind the thread around his fingers – either absent-mindedly or intentionally – which eventually became a small ball in his hands. When he bounced it on the floor, it rebounded high towards the ceiling. Haskell thought of the harsh gutty he usually played with and realised the wound rubber elastic band would be perfect as a golf ball if only he could cover it. When he mentioned the idea to Bertram, he suggested they use gutta-percha. The two immediately applied for a patent which was granted on 11 April 1899.

Initially handmade, demand quickly outran supply. The balls were rare in Britain and cost £1 each, well over £100 today. The 'Haskell ball' – perhaps 'Work ball' sounded wrong – quickly became the norm, especially after they had the bramble pattern moulded onto its surface. Within a few years, numerous patents were issued, all trying to improve on the wound golf ball. Eleazer Kempshall alone was awarded one golf patent every week between 1902 and 1904.

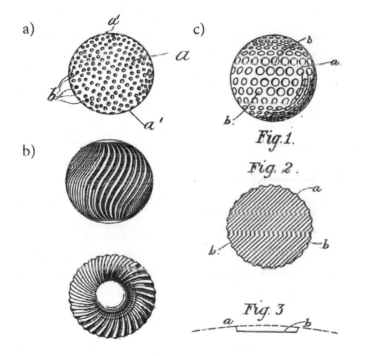

Figure 19. (a) Patent by Messrs Fernie (golfing professional), McHardy (Professor of Music) and Froy (manufacturer of stoves) of south-west London for an early prototype dimpled golf ball. (b) Patent by Alexander Henry of Edinburgh (gun and rifle maker) for a golf ball with curvilinear grooves intended to keep the ball straight when hit. (c) The first patent for a dimpled golf ball in 1905 by William Taylor of Leicester.

Patently obvious

It's in the nature of early patents, and no doubt some recent ones too, that some of the science is less than robust. Patent lawyers aren't necessarily there to check the science, only to determine who first had the idea, regardless of how crackpot it is. In golf, it seems, there were inventors far and wide hoping to make a fortune out of this lucrative new industry. They focused on the markings that had so much effect on the gutty, without really understanding why. The inventors in one patent sound like a bar room joke: three men walk into a patent office – a professional golfer, a music professor and a stove manufacturer. Their idea was to punch depressions into a smooth golf ball in what looked remarkably like regular dimples. Their claim? That it would make a rounder ball than one with bumps and would roll better. It was a good idea, but they'd missed the point: had they understood anything about aerodynamics they could have made a fortune.

Alexander Henry, a gun and rifle maker of Edinburgh at least understood something about flight. He filed a patent in 1898 for a golf ball with swirly grooves going from one pole of the ball to another. As a rifle maker, he was clearly influenced by the spiralling grooves down the inside of a gun's barrel which made bullets rotate in flight. Since bullets are always a little unbalanced, spinning them smoothed out any imbalances in the same way the ancient Greeks had done with their javelins thousands of years previously.

Studies on the flight of rotating bullets and shells had shown that rotation would make a bullet veer slightly to one side, requiring makers to alter the gun's sights to compensate. In 1852, Gustav Magnus did experiments in front of the Berlin Academy to show that the rotation of a cylinder in a wind tunnel made it swing to one side and, since then, the effect of spin causing a ball to deviate in flight has been known as the 'Magnus effect'.

Back in Edinburgh, Henry's bright idea was that his spiral grooves would make the ball have rifle spin like a bullet and keep the ball stable in flight. If he'd read any of Tait's papers, who was working just a couple of miles down the road, he'd have realised that the main spin is underspin rather than rifle spin. This rendered his invention useless.

There were hundreds of such patents on golf balls in the early 1900s; there are now tens of thousands. But one stood out due to the very precise nature of its claims. William Taylor, an engineer from Leicester, stated that his invention would 'obtain better results in the flight of the ball... giving a flat trajectory with a slight rising tendency, particularly towards the end of the flight'.[64] Taylor appeared to have designed a golf ball to optimise the trajectories that Professor Tait had so ably described.

Taylor said that rather than have the pimples of the popular bramble pattern, his new ball would invert the shape to have what we now call dimples. Taylor was extremely precise about what was required: cavities between nine and 15 hundredths of an inch in diameter (2.3 to 3.8 millimetres) covering a quarter to three quarters of the ball and no deeper than 14 thousandths of an inch (0.36 millimetres). Taylor was intriguingly particular about the shape of the cavities, saying they should be shallow but have a steep lip at their edges, a bit like craters on the moon. The steepness of the cavities was essential to the hanging flight, he said.

It was as if Taylor had a wind tunnel and test range, like manufacturers do today. Some speculated that he must even have been one of Tait's students. And yet Taylor was unknown to the golfing establishment. Just who was William Taylor?

He was born on 11 June 1865 in Hackney in what is now central London.[65] Taylor learnt to make things by watching the village blacksmith and wheelwright and would have seen Hackney turn from a village into a busy London borough after the arrival of the trams. He and his brother were fascinated not just by science but the art of making things: they made their own lathe, studied joinery, made their own telephones and even a phonograph. He became interested in lenses, magnifying glasses and telescopes. After an apprenticeship making scientific instruments, Taylor used his skills in instrument manufacture to make the sort of magic lanterns that Muybridge was using at the time to astonish audiences down the road at the Royal Institution.

By 1901, he had joined his brother's firm in Leicester and was married with four daughters under the age of eight. Evidently, he began to suffer from stress and his doctor advised him to take up a hobby: he chose golf. No sooner was he playing than he began thinking about the bewildering variety of golf ball designs and how to make them better. Like Ely Callaway 100 years later, he also had a catchphrase: "Never waste time in making what other people make; devise something new they've not thought of." He set off to find what the myriad designs had missed and his hobby of golf rapidly turned into a research project in the workshop he rented at nearby Narborough Hall.

Over in Paris, Étienne-Jules Marey had moved on from biomechanics to aerodynamics. The concept of human flight had been mooted for some time and he designed and built a wind tunnel to allow him to visualise and explain airflow over wings. His idea was very simple: rather than fly the object through the air and try to take measurements, he kept it fixed and blew air over the object. To see where the air went, he introduced smoke from 57 separate narrow pipes and sandwiched the airflow between two plates with the front one made of glass. Marey described his wind tunnel in an article in *La Nature* in September 1901 in which he ended with the following: "One may easily conceive the multiplicity of problems that may be solved by this method. We have described it in detail, so that it may be used by all…"

Taylor may have read this, because not long after the *La Nature* article, Taylor built his own glass-fronted wind tunnel to pass smoke over different golf balls. Once he'd chosen a ball design, he developed a device to machine moulds, made the moulds and then made the balls. Taylor tested the balls using an ingenious ball-driving machine he designed and built himself. It had a base one metre square so that it sat level on the ground, supporting a pyramid frame about a metre high. The point of the pyramid supported an angled axle holding a golf club so that it could rotate along one triangular face like the hand of a not-quite-vertical clock. When the club was drawn back, it pulled up a weight on a cable until it was stopped by a catch; pulling a trigger released the catch, the weight dropped and drove the club around the clock face to hit the golf ball sitting on a tee.

This ingenious contraption created the repeatable golf shot that Taylor needed to test his different ball designs. In a nearby field he studiously measured how far each ball went. By 26 April 1906, he'd written, submitted and been awarded a UK patent. His wife Esther supplied the word 'dimple' for the pattern.

The first licensee was Spalding who called it the 'dimple golf ball'. It came at a hefty premium of $9 a dozen – almost double the price of an average ball. By the beginning of the First World War, Spalding had sold 60,000 dimple golf balls, worth about $13 million today. The design was so good that after the war when the patent ran out, nearly every manufacturer used 336 Taylor dimples on their balls.

This is when Taylor vanishes from golf. His company's lenses had been in high demand during the First World War, used in aerial reconnaissance, binoculars, range finders and gun sights, for which he was given an OBE. In the meantime, his golf ball patent had run out and the golf industry went on without him. He continued to make the best lenses in the world and, in a rapidly expanding Hollywood, most stars stood in front of one of Taylor's precision-made lenses. He realised that the only way to keep them the best was to check and measure them during the manufacturing process. He developed a surface measuring machine called a Talysurf, a device that would draw a probe across the lens and measure its minute roughness down to millionths of an inch. Taylor reached the pinnacle of his career when he became the President of the Institution of Mechanical Engineers in 1932, although any mention of his golfing credentials is strangely absent from their archives.

Taylor's legacy reaches outside golf. Whenever a roughness pattern is used to change the aerodynamics of an object –

particularly in sport – it's usually said that it works 'like the dimples on a golf ball'. Taylor was the first to understand what was going on and the first to exploit it.

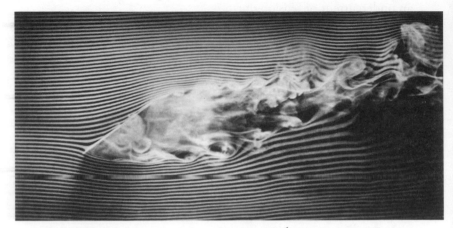

Figure 20. Image from an early wind tunnel developed by Étienne-Jules Marey. The image shows a plate at 30 degrees to the airflow entering from the left. © *Cinémathèque Française*.

Ruling the greens

Every manufacturer wanted their ball to fly farthest, or at least they wanted to claim it did. They tightened the ball's windings to make it stiffer and give it a higher coefficient of restitution off the club. They played with the size and weight of the ball. At one extreme was a ball that was both small and heavy: this had low drag, was relatively unaffected by gusts of wind and tended to carry far down the fairway. Being heavy, it needed energy to launch it, so was favoured by men. At the other extreme was a large and light ball. This would hang in the air and could even float on water, good for those who habitually found the water hazards. Britain's windy links courses suited the smaller ball; America preferred the bigger ball.

The Haskell ball and Taylor's dimples had added around 25 yards to a drive and now every ball that came out claimed to fly at least an extra five yards. The USGA and the R&A started to get worried: golf balls would fly too far, golf courses would become too short and the game would lose its appeal. The USGA called a conference with the R&A in 1920. Feelings were running high and, even when I started my PhD 65 years later, I felt there was still an undercurrent of tension during meetings between the two. The R&A still favoured a small ball, the USGA a larger one. Somehow the smaller ball was chosen. It would have a diameter of 1.62 inches and a weight of 1.62 ounces (41 millimetres and 46 grams).

Dissatisfaction at the USGA grew over the next decade until 1931 when the USGA decided unilaterally that America would play with a bigger ball. For the next four decades, two different balls were used across the world. They weighed the same but the American ball had a diameter of 1.68 inches, four per cent bigger and with eight per cent more drag.

The USGA put an initial velocity rule in place which said that no ball should come off a standard club at more than 250 feet per second, a little higher than found in P.G. Tait's experiments of the 1890s. By 1968, the R&A had changed their minds on the size of the ball. There were plenty of courses across the world that weren't like Scottish links courses and, anyway, the British were getting trounced when they played in America. Perhaps it was time to change. The bigger ball was introduced to all tournaments except the Open Championship, which managed to hold out until 1974.

But the battle between the ruling bodies and the manufacturers went on and an overall distance standard was introduced in 1976 which limited the distance the ball could travel to 271.4 metres. This included the trajectory, the bounce and the roll. The test used a golf swing robot called the Iron Byron, originally made by the True Temper shaft company. The robot was not so dissimilar to the device made by William Taylor in the early 1900s but had a joint for both the elbow and the wrist.

It was mothballed in 2002 when it was replaced by a much more sophisticated indoor test range.[66] This comprised two machines. The first fired golf balls using a mechanical tester to get a ball's initial launch conditions for a standard shot in terms of speed, spin and angle off the club face. The ball was then fired through a set of rotating wheels quite like the ball-firing machine I used to fire golf balls onto golf greens. The counter-rotating wheels would give the ball a range of speeds and spins as it was fired down a 25-metre tunnel to measure the ball's lift and drag characteristics.

The launch conditions and the aerodynamic characteristics were enough to create a virtual trajectory of the ball landing on a virtual fairway and its subsequent bounce and roll. I used the same technique in my PhD and in Tennis GUT. The simplest impact model between the ball and the turf simulates it as a spring and damper, similar to the way McMahon and Greene modelled the interaction between the foot and running surface in Chapter 1. The USGA either did something similar or, more likely, did an array of tests on real turf so that they could put in landing speed, spin and angle and look up the actual bounce and roll from a table. The new rule for the total distance a ball could travel was increased to 320 yards (292.6 metres).

By 2002, the rules fixed the size and weight of the ball, its maximum launch speed and the maximum distance it could travel. The only way the manufacturers could differentiate themselves from each other was the trajectory between the ball impact with the club and it landing on the turf. It was time to return to the dimple.

Crisis

Until the 1970s, most golf balls had 336 Taylor dimples in an Atti pattern, named after the prolific mould maker Ralph Atti of New Jersey. Surprisingly, the first mention of the word 'dimple' in any patent was as late as 1970. It was submitted by Acushnet and claimed a three- to five-yard advantage for dimples with a new saucer-shaped series of indentations. Other patents rapidly followed; the Titleist icosahedron pattern filled the ball with dimples arranged into triangles, while the High Velocity Core ball had five different dimple sizes on its surface. Dimples were made bigger, smaller, deeper, shallower, further apart and closer together. Almost any design was used, all accompanied with the claim of a three- to five-yard advantage.

Manufacturers copied Taylor's approach by testing them out on the firing range. Differences in the size of the dimple were now down to hundredths of an inch, so it was important that balls were measured accurately when they came out of the mould. Often, they used machines like the Talysurf to do this, not realising that it was invented by the same person who'd invented the dimple itself.

The aerodynamics of a golf ball has consumed an awful lot of time in wind tunnels across the world, which are now huge devices taking up whole rooms. The ball is held in the airflow from behind by a thin stiff horizontal bar called a sting which takes a right angle turn down to a force platform below to measure the forces of the air pushing on the ball.

All studies show that drag force is proportional to a ball's cross-sectional area and its 'drag coefficient'; this is known as C_d in the industry (pronounced see-dee). C_d is a beautifully simple number and its value can give a clue to the way the air is flowing over the body. A typically aerodynamic object like an aeroplane wing has a low C_d of 0.04; a cyclist sitting upright on a bike with long flowing hair has a value of 1.1. A smooth ball is in the middle, at about 0.5.

At very low wind speeds, the airflow across a smooth ball is laminar since it flows over the surface in neat layers one on top of the other. The layer at the surface is stationary, while others further away slide over the top of it. Think of a thick ream of writing paper sitting on a table: if you rest your hand on the top and slide it forwards, the top layers move easily while the one at the bottom in contact with the table doesn't move at all. Aerodynamicists call the slow portion in contact with the surface the 'boundary layer'; in air it is only tenths of a millimetre thick. It is the source of everything good and bad about the flight of a golf ball.

While air flowing at low speeds is laminar, at high speeds it becomes turbulent. You can see this effect for yourself on another fluid if you run a tap into your sink. Run it slowly and the water comes out in a nice smooth pipe of water; gradually open up the tap so that the flow increases in speed and the smooth water turns ragged and turbulent. A stream of a fluid will always transition from laminar to turbulent flow if the speed is high enough.

When air hits a ball, it strikes the front and is pushed around the sides. The air in the boundary layer in contact with the ball slows as it goes around it and, eventually, is so slow that it detaches from the surface of the ball. The air leaves the ball near the poles and the wake is large: this indicates a large drag force.

Increase the wind speed and the air in the boundary layer eventually turns turbulent. The effect of this is to mix high speed air away from the ball with low speed air close to it. Since the boundary layer is now just a little faster, it travels further around the back of the ball, the size of the wake is reduced and so is the drag. Aerodynamicists call this transition from laminar to turbulent flow and high to low drag, the 'drag crisis'.

The speed that the drag crisis occurs at for a smooth ball is very high, much higher than someone can hit it. A roughened surface, however, brings the drag crisis speed down right into the range of possible golf shots. This is exactly what happened with the gutty. At the beginning of the day, the ball was smooth and the drag coefficient was high at around 0.5; as the ball got roughened up, the boundary layer became turbulent, it stayed attached to the ball further around the back and the wake became smaller. The consequence was that the drag coefficient dropped to around 0.2 and the ball flew further.

Taylor must have seen this effect in his wind tunnel and used the dimple to force the drag crisis to occur at low speeds. It was the rear lip of the dimple crater that was key: if it was just the right steepness, it would trip the laminar boundary layer into turbulence without making it bounce off the surface and separate from the ball. By the 1990s, computational fluid dynamics, computer-aided design and rapid manufacture meant that designing dimples was relatively quick. Manufacturers investigated every possible dimple geometry and pattern to give flights and trajectories to suit different styles of player.

The different roughnesses can be represented as a series of peaks and valleys.[67] Close up, the dimples of golf balls are like a series of long shallow valleys with narrow flat peaks between them. The original bramble pattern of the early gutties is the inverse: very narrow valleys with wide plateaus between them. The essence of low drag is to keep the boundary layer energised. The lip of the dimple crater does this, whereas the long smooth plateaus of the bramble just slow it down again (as do the panels on a football). The upshot is that if low drag is the desired outcome, dimples are better than pimples and pimples are better than a smooth surface.

Many club manufacturers copied Callaway by making larger-headed clubs. But, as the club heads became larger, our aerodynamicist, Dr John Hart, realised that drag had become more important. He put ridges on the top of the club head called riblets; these worked just like dimples by creating turbulence in the boundary layer as air flowed across the top of the club. The result was reduced drag and increased swing speed. We shared the patent with Ping and they incorporated the idea into their G25 driver – it became their best-selling driver of all time.

Golf is a superb example of innovation by manufacturers and the counter-measures needed by the USGA and R&A to make sure that the game isn't ruined. The biggest fear is that driving distances will increase so much that golf courses will become too small and golfers will get bored and leave the game. That wouldn't help anybody in the golf industry. The ruling bodies regularly put out a review of driving distances. In 2017, the report noted that driving distances of male amateur golfers had grown by four per cent over 21 years to 190 metres. The top ten professionals were driving so well in tournaments that the ball was stopping about eight metres short of the absolute limit. It seems likely that the current rules are keeping overall ball distances in check for the moment.

The report concluded that the period between 2004 and 2016 was characterised by a period of 'stability through regulation' but also noted that 2017 showed a significant rise in driving distances of two to three metres.[68] This may be down to a random fluctuation due to good weather conditions or a particularly good set of players within the year. However, the implication coming out of every paragraph of the report is: 'we have rules and we're not afraid to use them'.

The following chapter describes the next sport in our journey through time. It is one of the most technological of all sports and shows how aerodynamic features as illustrated by the golf ball dimple can improve performance dramatically, particularly if you're travelling at over 100 kilometres per hour.

SEVEN

Sliders

I CHANGED trains at Chur, a pretty market town in a flat valley 190 kilometres south-east of Zurich. My phone told me it was only 50 kilometres to my final destination, St Moritz, and yet it would take another two hours: that was 25 kilometres an hour. This was hardly an express. What was wrong with the train? Was it broken? I shouldn't have been so churlish; the train chugged up a steep gorge, past the snow line and into the mountains, gliding over high viaducts as green-glass rivers slipped by far below. Waterfalls were frozen in time as the snow on the ground thickened and the train travelled further up towards the summit at just under 2,000 metres.

I was on the way to St Moritz to do some experiments for a second Kensington TV documentary on the Winter Olympics.[69] I got a sudden glimpse into the origins of what I was here to research: two sleds bearing people with thick coats and even thicker grins flashed past beneath the viaduct and vanished down a snaking snow-covered road. This is how it must have been a century ago for the rich Victorian tourists who came this way to take part in what is surely one of the strangest and rarest of sports – bobsleigh.

Most people will have seen a modern bobsleigh on their television but probably not in person. There's a version for two people – women or men – and a version for four men only. There are also the more traditional-looking sleds of luge and skeleton; on skeleton, you go down head first, on the luge you go feet first.

In 1855, Johannes Badrutt and his wife Maria bought a small hotel in St Moritz. They renamed it the Kulm Hotel and started to expand it. In 1864, not satisfied with the short summer season, the Badrutts made their regular British guests an offer: come back and stay for the winter, and if they didn't like it they would reimburse their travel costs. The *quid pro quo* was that if they did like it, they should tell all their friends. The guests accepted, arriving at Christmas and staying until Easter, departing tanned and happy. Winter tourism prospered and the entertainments blossomed. Guests borrowed local delivery sleds called *slitterli* to have impromptu races in the streets.

The rival spa town of Davos 80 kilometres to the north hosted its own winter tourists and, by 1883, had held its first toboggan race, creating the Davos International Toboggan Club. This spurred on the guests in St Moritz and the Kulm Hotel's Outdoor Amusements Committee, comprised of five well-to-do British gentlemen. The committee decided that they would have their own sled race from St Moritz down through the fields to Cresta. This would have the added advantage that it got them off the roads of St Moritz where they were annoying the locals.

The Badrutts, keen to keep their guests entertained, had their workers handcraft a path through the snow with challenging banks and turns and named it the Cresta Run. It opened on New Year's Day 1885. Sliders initially went feet first, sitting almost upright with a slight backward lean. An American tourist introduced a low-slung sled made by Allen and Company in Philadelphia. Called the Flexible Flyer, it had a T-bar at the front attached to the runners; pushing on the bar flexed the runners and caused the sled to turn. The T-bar could be used with the feet if sitting upright, but could also be used with the hands if going down head first: eventually they all adopted this position.

Figure 21. A Flexible Flyer sled from 1936. *Picture courtesy of The Children's Museum of Indianapolis.*

The St Moritz Toboggan Club was founded two seasons later in 1887 with its first committee consisting of Mrs Shepley, Miss Cousins and Messrs Topham, Cremers and Watson. By 1888, longer 'bob-sleds' were starting to appear. There are two arguments about the history of the bobsleigh: firstly, who invented it; secondly, the origin of the name. The International Olympic Committee says that the Swiss created the winter sport of bobsleigh in the 1860s; the *Times Union* newspaper claims that the birthplace of bobsleigh was Albany, New York in the 1880s; the International Bobsleigh and Skeleton Federation evades the topic completely and simply mentions the first bobsleigh club in St Moritz. Older references talk about a British inventor called Wilson Smith who was supposed to have given the name 'bob-sled' because the riders would bob back and forth to get it going.

Max Triet set the record straight in *100 Jahre Bobsport*, a book and exhibition to celebrate the centenary of bobsleigh in 1990.[70]

Triet was unequivocal: the term bob-sled was used as far back as 1839 to describe transport used by lumberjacks for carrying logs and timber out of the forest. These had two sleds attached together so that the front one pulled the rear one as a trailer, much like lorries on the road do today. When the trailer is removed from any sort of tractor, the foreshortened front part is known as a bobtail and it seems that the sleds then became known as bob-sleds.

By 1885, Albany was holding multi-person bobsleigh races during their winter carnivals and one of their residents, Stephen Whitney, went to Davos on holiday. In December 1888, he took two sleds and attached them together with a long plank of wood so that it would fit more people. A St Moritz blacksmith called Christian Mathis copied the design and made a multi-person sled with steering and brakes. While the Cresta Run had really been designed for single-person sleds, by the end of the 1891–92 season, they hosted the first bobsleigh race for teams. Five bobsleighs took part in the race on 19 March 1892. The team in second place had as its members Wilson, Smyth, Duff and Jones; the first two names in the team probably gave rise to the myth for many years that Wilson Smith was the inventor of the bobsleigh.

There must have been tension between the single riders on their low-slung sleds and the teams on their larger bobsleighs because, by 1903, a gala was held to collect donations for a new bobsleigh track to take the bigger sleds off the Cresta Run. They collected nearly 11,000 Swiss francs and construction soon started on a route from the grounds of the Kulm Hotel down to Celerina below. The first races were held on New Year's Day 1904 and St Moritz has been synonymous with the bobsleigh ever since.

The two tracks are made from natural snow and ice, rebuilt each year from that season's snowfall. The work is so specialised that the task has been carried out by generations of the same family. If you visit in summer, the tracks are like the remains of melted ice lollies with only the sticks left; all you can see are the vague outlines of the track with the buildings at each end. After the first big snowfalls in November, workers from South Tyrol begin to shovel snow by hand, trying to recreate the same track each year, literally sculpting their way down the hill. The Olympia Bob Run St Moritz-Celerina uses over 20,000 cubic metres of snow and water to drop a sled down a one-in-seven slope from St Moritz at the top to Celerina in the valley below. It's not just a straight line, of course; there are plenty of curves introduced for the early British tourists. These have inventive names like Sunny Corner, the Horseshoe and Devils Dyke, becoming progressively more punishing as you go down the hill. The Horseshoe subjects sliders to over four gravities of acceleration as they navigate the curve.[71] In contrast, astronauts taking off in the space shuttle feel an acceleration of only three gravities, (although for a much longer time).

Given the history of St Moritz, I wasn't surprised to find a museum of the bobsleigh. It's an eclectic mix of paraphernalia going back to the late 1800s and is the pride and joy of Donald Holstein, a Swiss bobsleigh pilot (the person at the front doing the steering). As with most small towns, many people have more than one job and Donald looks after the museum, is vice president of Celerina, runs the local bike shop and even takes paying tourists down the bobsleigh track. The tour of the museum is very hands-on without the 'do not touch' signs usually associated with museums; as a man who must have sat in a bobsleigh many thousands of times, Donald views the sleds as living objects rather than sterile museum pieces. This was good, because I was desperate to sit in one.

Leaning against a wall was a double-length sled with the name 'Mathis' imprinted on it.

"People think the bob word comes from the sliders bobbing up and down during the run to help them go faster, but this is not true," said Donald. "It is after the name of the sleds the workmen towed behind them in the forest to bring out logs and trees." Donald knew his stuff; I nodded wisely in agreement.

As befitting a blacksmith's construction, Mathis' sled was steered by two chains with hand-sized links at their end, like something you might see attached to the nose of a bull. Steering was done with a simple left or right pull of the chain to point the front section in the direction you wanted to go. The person at the rear had a handbrake lever either side attached to the end of a rake, so that pulling the levers up pushed the rake down into the snow to act like a plough. Rustic and effective.

The next bobsleigh Donald showed me was also double length but upholstered with a thick floral-patterned cushion; this could easily have been a bench in the foyer of the Kulm Hotel had its real purpose not been given away by the runners underneath and the massive bumper bar at the front. A grainy black and white photograph next to it showed one of the many Crown princes who seemed to have visited St Moritz sitting astride it.

Steering wheels appeared on motor cars in 1898 and seem to have made their way onto bobsleighs soon afterwards. An unusual bobsleigh variation was for five, where the team all lay down head first, layered one on top of each other like a fallen cascade of dominoes. The stem of the steering wheel was short so that the driver could rest on his or her elbows while still steering and not bang their chin on it. A picture showed a happy group of laughing men and women in just such a position. It was considered a little risqué, said Donald, and was eventually restricted to men only.

The *Feierabend* bobsleigh became popular between the First and Second World Wars. These were all steel and had a short fairing to cover the feet. The runners had a U-shaped profile as they gave better traction than the then traditional round bars and the rear runners were given axles with ball bearings to allow them to follow the contours of the ice.

By the late 1940s, the fairings had become elongated like the bonnet of a car covering the full length of the legs of the driver and the steering wheel. Undoubtedly, this would have improved the aerodynamics but Donald suggested that it was at least as much to do with creating a space to put on a sponsor's logo – bobsleigh was, and still is, an expensive sport. The seating was made of interwoven straps of hemp, a bit like a garden seat, hanging disconcertingly from the frame of the sled. I imagined what it would be like to dangle my backside in it with nothing between me and the ice as it hurtled down the slope at over 100 kilometres per hour.

The bobsleighs we see today look much more solid than the 1940s versions, with a fairing that stretches from the bulbous nose all the way back to the brake person. There are small foam seats and the floorpan is reassuringly solid. The steering wheel vanished after the 1940s, reverting to the original design of the Mathis sled with a cable and pulley on either side to pull left or right. The brake remains as rustic as ever, with the same two levers to pull upwards and a rake-like claw underneath to act like a plough.

Pushing the limits

The track had to be adapted as the performance of the athletes and their bobsleighs improved over the years. The most memorable turn at St Moritz is still probably the Horseshoe, as it brings the sliders a full 180 degrees back on themselves. The sled enters the turn on the flat ice, but slides up the ice wall as it goes around the bend so that it is pressed vertically up against the icy bank by the acceleration. By the 1950s, speeds had reached 140 kilometres per hour and the Horseshoe had a permanent vertical stone wall fitted for the temporary ice wall to sit against. Further down the track, speeds increased to 160 kilometres per hour and the old finish in the field was no longer enough to stop them. Instead, the track was doubled back on itself another 100 metres or so, back up the hill to a new finishing lodge.

There are strict rules on bobsleigh design that evolved in tandem with the sport. Firstly, a weight limit was imposed; the maximum for the two-man bobsleigh is now 170 kilograms when empty, and 390 kilograms when occupied.[72] It's not immediately obvious why there should be a limit to weight since Galileo showed over 400 years ago that all objects accelerate at the same rate regardless of mass when he dropped two spheres of different weights off the Leaning Tower of Pisa.[73] It would be logical, then, that the weight of the bobsleigh would be irrelevant. The problem with Galileo's experiment is that he didn't take into account drag. If he'd dropped a cannonball and a much lighter inflated ball the same size, the cannonball would easily have hit the ground first. The upward drag on the inflated ball would have been large in comparison to the downward gravitational force which would have slowed it up. The experiment is only really valid in a vacuum or where drag is negligible. For a sport like bobsleigh, where winning margins are small, then drag becomes important and so does the weight of the sled.

One organisation that knows the rules of bobsleigh well is the Institut für Forschung und Entwicklung von Sportgeräten (Institute for Research and Development of Sports Equipment or FES for short). Based in Berlin, it was founded in 1962 in the German Democratic Republic to create technologies that would enhance the performance of the East German team. In the late 1970s, the team burst onto the scene using bobsleighs with encased sides, rubber fixings and push bars. In Sarajevo in 1984, the Soviet Union countered with 'cigar' bobsleighs that were very fast but difficult to steer.[74] They had suspension systems with shock absorbers allowing each runner to move independently but were so dangerous that crashes began to mount up.

While the ruling body wondered how to regulate the sleds, others were looking for safer ways to train and learn the tracks. Mont Hubbard, from the University of California, Davis, developed a virtual reality bobsleigh simulator to help the US Olympic team train. He used three-dimensional Newtonian principles to model the motion of a bobsleigh down a track. He'd measured the track and modelled it as a surface and used a computer program to control the motion of an indoor bobsleigh using servomotors. A screen at the front showed a simulation of the track in tandem with the motion of the bobsleigh which could move from side to side, just like a virtual reality ride at a fair. The athletes could learn the track without even setting foot outside the lab.[75]

Friction between the runners and the ice is supremely important: it depends upon the material and coating used for the runners, the temperature of the ice and the runners' shape. The trick is to get enough friction to steer with, but not enough to slow you down. On the skeleton and the luge, athletes use their bodies to twist the frame of the sled which distorts the runners just enough to steer in a similar way to the early Flexible Flyers of the 19th century.

I was external examiner in 2010 for a PhD student called James Roche from the University of Southampton. He was one of two PhD students working on the design of a skeleton bobsleigh for the British team, and used a similar mathematical approach to Hubbard to work out the best trajectory down a track, and the design of sled that would allow a slider to take it. In the skeleton, athletes push the sled at the start with one hand and then have to jump dramatically onto it head forwards. My team's job at that time was to shape the metal saddle on top of the sled to receive the athlete. It had to be tight enough to give the athlete good control of the sled but not so tight that he or she couldn't get in. Happily, Amy Williams won gold at the 2010 Vancouver Olympics.

Bobsleighs are now made with a front and rear part and a ball joint to allow the front section to twist in corners; this keeps the runners in contact with the ice as it follows the contours of the track. The friction of the runners depends upon the weight of the bobsleigh too. The heavier the sled, the more the runners dig into the ice and the higher the friction. One thing I'd never appreciated is that the friction increases in the bends because the high G-force pushes the runners even harder into the ice. Bobsleigh designers realised that having a little bit of lift kept these downward forces to a minimum and ice friction low. They had to be careful, however, as when a bobsleigh came off a steep turn it could rebound off the ice and literally take off. If this happened, there was no amount of steering that could control the bobsleigh while it was in the air.

Watching a bobsleigh coming around the 180-degree Horseshoe turn is spine-tingling. If you stand in the middle of the bend in the spectator zone, the wall of ice curves right around you. You first hear a low rumble to your right, slowly increasing in volume until a bobsleigh and its occupants appear high up on the ice in front of you, almost within touching distance. The after-image is still there as it vanishes to your left with a rattle as the bobsleigh thumps down onto the flat ice further down the track.

Every sport has its own ecosystem. At St Moritz, the track builders split up into zones and constantly groom their creation, scraping away gashes and brushing off snow. The bobsleigh manufacturers design their sleds with a mixture of engineering and art and bring them out of their trucks like they're delivering a baby. Finally, there are the athletes who work together to coax half a tonne of bobsleigh down an icy chute, avoiding the edges and trying to get the best trajectories they can. The smallest time difference measured is one hundredth of a second; at a typical top speed of 160 kilometres per hour this is a difference of just 40 centimetres at the finish. Taking the wrong line anywhere down the track makes all the difference.

The building of a natural ice channel to slide down is not easy and getting the banks and turns correct needs a good eye and perseverance. The season only lasts for three months and all the effort and energy put into the track melts away with the spring sun. At one time there were over 60 other natural tracks in Switzerland alone; now there are also concrete-reinforced artificial tracks across the world at places like Igls (pronounced eagles), Whistler and Sochi, all seeking to emulate the challenge of the original at St Moritz. These artificial tracks have doubled the length of the sliding season and increased the speed and technical difficulty of the runs, sometimes by too much.

St Moritz is dominated by snow, ice and slopes. I marvel at the audacity and arrogance of the Outdoor Amusements Committee of the Kulm Hotel to create the St Moritz Bobsleigh Club and, in their own words, 'to spend as much money as it considers appropriate for different people to use Cresta Street'.[76] Did the Badrutt family who did so much to promote it know this would create a winter sport that would be the mainstay of the Winter Olympics?

Old vs new

Modern bobsleighs are low and sleek in comparison to their earlier versions. They are narrow and long like a missile with just enough room to get the athletes inside. They have stubby aerodynamic wings fore and aft above the runners, but are not allowed to cover them. The rules are extremely strict: the top and rear of the bobsleigh must be open, everything must be fixed and can't be altered during a run, no additional aerodynamic features such as holes, fairings or vortex generators are allowed; and the runners must be made of a prescribed material with no coatings.

Aerodynamic drag is directly proportional to the cross-sectional area of the bobsleigh, so making it smaller and narrower gives a distinct advantage. The rules, however, limit just how small the bobsleigh can be. They even specify the shape of all the curves across the bobsleigh so that the nose has to bulge outwards.

Figure 22. Top and left: Max Arndt pushing off a 1940s sled with his brakeman Alex Roediger at St Moritz. Right: Going around the Horseshoe turn. © *Kensington Communications*.

The TV producers I was working with weaved their magic and arranged for two German sliders to compare a modern bobsleigh with one from the 1940s. The two sliders, Max Arndt and Alex Roediger had 13 medals between them, including a 2013 World Championships gold in the four-man bob when it was held in St Moritz.

First, they did a run in their normal modern bobsleigh. I waited at the finish, watching the live screen above my head. They were dressed in tight black Lycra, spiked sprint shoes and sleek helmets adorned with the German flag. Max leaned into the cockpit of the bobsleigh to pull a lever and a push handle popped out at the front inviting him to push. It started to snow quite heavily and one of the Tyrolean trackmen cleared the start with a brush that looked remarkably like a witch's broom. Max and Alex pushed the bobsleigh gently to the start where they rocked backwards and forwards slightly to establish a rhythm. Suddenly, they were off, pushing the 170-kilogram bobsleigh as hard as they could. Max jumped in over the side and Alex slid in behind him, their arms disappearing fluidly into the bob like an octopus vanishing into a hole until all that could be seen was the tops of their heads. The front push handle popped back into the body.

Their time to the 50-metre mark flashed onto the screen: 5.2 seconds. This compared well to the recent World Cup times at St Moritz. The screen cut automatically to the different turns, some of the cameras obscured by the heavy snow – Sunny Corner, the Horseshoe, Devils Dyke, Bridge and down to the maximum speed point just before Martineau. Their top speed came up: just under 134 kilometres per hour with a finishing time of 1 minute 11.80 seconds.

The bobsleigh scraped to a halt in front of me as Alex applied the brake, clawing parallel lines behind it in the snow.

"I couldn't see the line," said Max's muffled voice from behind his helmet. "Too much snow." The trackmen manhandled the bobsleigh, Max and Alex into a small truck and they trundled back up the hill to the start.

The producers had magically sourced a 1940s sled and clothing from the era. The Lycra suits were replaced with thick woollen jumpers and heavy trousers, the helmet was made of leather and had separate goggles, and the shoes were heavy military boots with thick steel studs in their soles. The difference between the old and modern bobsleigh was stark. Gone was the sleek carbon fibre body with the stubby aerodynamic wings. Instead, here was a simple chassis with a thick bulbous front and a seat made of hemp.

I watched them appear at the starting position on the big screen; the bobsleigh looked strange in comparison to the modern sled, almost like the bones of a large exotic fish. Alex took his place at the rear, holding on to two vertical posts attached to the chassis while Max leaned over the cockpit to push on the front. It certainly looked less efficient than the modern bobsleigh. They slapped their bodies to shake off the snow and set off at a slow sprint. Their boots were thick with ice and they slewed visibly. They jumped in easily through the open body.

Their start time flashed up on the screen: 8.24 seconds. Max had mentioned that whatever you lost at the top in time was tripled at the finish. At this rate, they could be nine seconds down on the modern bobsleigh. They negotiated Sunny Corner, then the Horseshoe. Their maximum speed and finishing time came on the screen: 108 kilometres per hour and 1 minute 27.01 seconds, 16 seconds down on the modern bobsleigh.

I looked down the track, expecting them to appear up the hill. Nothing. Had they crashed on the final corner? Unperturbed, the Tyrolean track marshal stomped down the track, unreeling a cable behind him. Another marshal switched on a winch and the cable went taut. I looked down the track and the Tyrolean's wide-brimmed hat appeared over the brow of the incline. More and more of him appeared until I realised he was standing on the rear of the bobsleigh which was being pulled up the track by the cable. Max and Alex raised their arms in triumph.

I realised my mistake. Of course they were OK – there was no crash. I'd forgotten that the 1940s bobsleigh was so much slower that it hadn't managed to get up the hill. They'd stopped exactly where they used to 80 years ago.

Max and Alex jumped in their truck and I raced up the hill after them to meet them in the café overlooking the start in the fantastically named Dracula Start House. It has vintage winter sports paraphernalia hanging from the ceiling and the smell of glühwein in the air. It's pretty cosy and just a couple of dozen people can fill it packed around the circular bar, coats and winter gear steaming gently on the windowsills.

I wondered how much of piloting a bobsleigh was psychological and how much was physical. I'd given Max a heart rate monitor to gauge his response to the two different bobsleighs. In the modern bobsleigh, his heart rate at the start line had been a low 70 beats per minute, equivalent to a very gentle walk. It shot up during the push start and peaked at the Horseshoe turn at about 120 beats per minute before dropping back to 70 at the finish.

I asked him what it was like with the old bobsleigh.

"I felt nervous at the start," he said. "I've never been in an oldie-bob before."

The heart rate monitor bore this out: on the start line, his heart rate jumped to 130 beats per minute, higher than at any point with the modern sled. What happened next showed what a true professional Max was. As soon as he was in the bobsleigh, his heart rate dropped to 80, increasing only to 94 at the Horseshoe and dropping back to just over 70 even before the finish. Max explained that as soon as he realised the bobsleigh steering was OK, he relaxed and enjoyed himself. The lower speed meant that the Horseshoe wasn't so intimidating and his heart rate remained low. Only an Olympic athlete can relax doing four gravities of acceleration vertically around a bend at 100 kilometres per hour.

On the day, in the same snowy conditions, the new bobsleigh outperformed the old sled by around 16 seconds, a 17 per cent improvement. About three per cent was due to the poor start and 14 per cent due to the run itself. I created a simple mathematical model like that of Hubbard and Roche to give me an idea of where the 14 per cent might have come from.[77]

In the old bobsleigh, Max and Alex had sat upright, exposed to the oncoming air. This would have made their cross-sectional area really quite large. The cigar-like tube of the modern bobsleigh on the other hand would have kept it to a minimum. The sides of the old bobsleigh were completely open and this would have increased the drag coefficient, as would their rough woollen jumpers and old-style helmets. The modern sled would have decreased the drag coefficient with its filled-in sides and smooth finish.

My model estimated that the better bobsleigh design reduced the drag by two thirds. This allowed Max and Alex to travel 23 per cent faster in their new sled, giving them a 16-second advantage. If they'd been side by side at the start, the modern sled would have been 400 metres in front at the finish.[78]

Dead weight

It seems that St Moritz hasn't changed much since the Badrutt family worked to make it the winter tourism capital of the world. People still flock there to take in the beautiful winter air, although staying from Christmas to Easter at the Kulm Hotel would now cost you more than the original hotel cost to build in the first place. The night before I left St Moritz, Donald Holstein, the owner of the Bobmuseum, asked if I'd like to be his brakeman the next morning as he was taking some tourists down the St Moritz-Celerina track in a four-man bob. I quickly agreed.

I had a terrible night's sleep, dreaming repeatedly of pushing the bobsleigh but failing to get in and watching it slide off into the distance without me. The next morning, I walked wearily to the start, chose a helmet and jacket, doing my best to look like a professional brakeman and waiting for Donald to tell me what to do.

"Just get in the back after the tourists and someone will push us off. Don't worry about braking, I have a footbrake."

I'd never been so relieved to find out that my sole task was to be a dead weight: it was going to be fun. We pushed the sled onto the start as I'd seen Max and Alex do, had our photos taken and

I clambered in, Donald as the pilot and the two tourists between us, while I did my best brakeman act. It was quite a squeeze and I had to reach forward to find the hand-holds in the floor, my elbows resting against the metal ribs of the chassis. The inside wasn't as tidy as I'd expected, with metal ribs and bolts jutting into my body. The thin cushion I sat on slid around worryingly. Someone pushed us off and we rattled disconcertingly down the tight U-shaped duct of ice, the noise of the wind rising past my helmet. I saw Sunny Corner appearing rapidly ahead as a wall of ice going around to the right.

I couldn't wait to see what it looked like from up on the ice. We entered the bend and the sled tilted sideways up to the left until we were vertical against the left-hand wall of ice. Before I had a chance to take in the view, a massive force pushed my head down between my knees: I saw nothing but the dirty belly-pan of the sled. I'd forgotten about the huge accelerations of the curve.

My head rebounded upwards just in time to see us enter the left-hand bend of the Horseshoe turn and my head was thrust down again. This happened again with varying degrees of force on each subsequent bend until I heard the clawing of the brake underneath my backside and Donald brought the sled expertly to a stop.

I got out exhilarated and slightly dazed. I congratulated the tourists and asked them if they'd enjoyed it, pushed the sled into the back of a van and we were transported back to the car park at the start. I realised that my arms and neck hurt but, hey, I'd just been down the St Moritz-Celerina track at over 100 kilometres per hour. I could tick that off the bucket list and never have to do it again.

At the top, I was a little confused as Donald pushed the sled back to the start. "Again?" I asked, bemused. "Again," said Donald.

We repeated the run with a different set of tourists. I desperately tried to keep my head up so that I could see the turns but it was impossible to fight the four gravities of acceleration. If a head weighs about five kilograms, then effectively it weighed 20 kilograms going around the Horseshoe. No wonder I couldn't hold it up. After the second run, I felt a little light-headed and slightly nauseous. I asked Donald nervously if we were going again, suspecting I knew the answer already. "Yes," he said, "just the four times."

I hung on, smiled and whimpered through the next two runs. The nausea took the rest of the day to clear, the bruises to my arms and legs faded after three days and my bad neck lasted a week. I look at the likes of Max and Alex with new-found admiration: day after day of punishing discomfort with every chance you might not win. That takes dedication and talent and the best engineering an athlete can buy. And you have to be slightly barmy.

In the next chapter, we approach the present day and find that athletes in even the simplest of sports will go to great lengths to win. In the next sport, technology got a little out of hand and we find out what happens when a sport allows technology to develop unchecked.

EIGHT

A leap into the unknown

PAUL BIEDERMANN and I stood by the side of a 1960s pool in Dresden in former East Germany, trying to squeeze his muscular frame into what was effectively a sheet of stiff polyurethane glued into the shape of a swimsuit. It was jet black and extended from his ankles all the way up to his neck, leaving just his shoulders and arms bare. He'd put plastic bags on his feet to help slide them through the leg holes and it'd taken him 20 minutes to roll the suit up past his trunks. He'd pulled the straps over his chest and needed help to zip up the back. Some have likened it to making a human sausage.

I carefully pulled the zip up between his shoulder blades, the Y-shape straining as it reached the top. In the heat of the pool, I felt a bead of sweat trickle down my back as I carefully inched it closed; this was the moment where the $300 X-Glide suit could split embarrassingly. Paul was – and still is at the time of writing – the world record holder in the 200 and 400 metres freestyle. 'Freestyle' means you can use any type of stroke to get the fastest time; in practice, this means the front crawl.

Paul's name might have come into your consciousness in 2009 when he had the temerity to beat Michael Phelps in the 200 metres final at the World Championships in Rome. His world record time of one minute 42 seconds took almost a second off Phelps' record the previous year, beating him by over two metres.

In total, 43 records tumbled during the championships in Rome; eleven of them were broken twice in the same event. What was going on? World records were becoming worthless.

The cause for the record bonanza was obvious, said the critics: it was no longer about the athlete, it was all about the swimsuit. Technology was advancing too fast and the swimmers had been lost in the midst of it all, they complained. In the press conference after the race, Phelps was asked: "Were you beaten by the swimmer or by the technology?"

Graciously, he responded that it was the swimmer, that it was his fault alone and he'd prepared badly. His coach was a little more forthright: "The sport is a shambles right now and they'd better do something or they're going to lose their guy who fills these seats." It was a clear threat to the governing body: do something or Phelps wouldn't swim again.

The months leading up to the Championships had been frenzied as athletes had desperately investigated the rapidly changing suits. Which suit should they choose? What if the best suit was not their sponsor's? They didn't want their rivals to have an advantage but some like Rebecca Adlington, British double gold medallist in Beijing, refused to change her suit, claiming that the new ones were tantamount to technical doping.

How did the suits work? Was it doping? Should they be banned? All eyes turned to the ruling body FINA to do something.

The million-dollar mermaid

Of course, it hadn't always been like this. Go back far enough and you find people swimming without swimsuits at all. The ancient Greeks and Romans loved their baths and spas and knew that swimming was good for physical training; there is even evidence of swimming events being held in the bay off Athens.

In the Middle Ages, swimming was frowned upon because it was thought to spread disease, but it returned during the 17th century as people flocked to spas and warm springs. The question of what to wear became important, especially during mixed bathing. Women were encouraged to wear stiff canvas gowns that wouldn't cling to their bodies and would hide their shape. Men wore drawers and a waistcoat.

Swimming as a pastime didn't really take off until the middle of the 19th century, coinciding with the rise of leisure time, the expansion of the railways and holidays by the seaside. Nude bathing was banned in Britain by 1860 and the term 'swimming suit' was first used in the 1870s.[79] Early swimwear tended to reflect the underwear of the day so that the men's version resembled long-johns. Women's were more complicated, with a loose-fitting dress down to the ankle held down by weights in the hems to stop them floating upwards in the water. Early film footage, possibly from the first Olympics in 1896, shows women wearing a hat, long-sleeved blouse, skirt to the ankles, leggings and plimsolls diving fully clothed into the water. They looked nervous and uncertain in front of a bemused crowd.[80]

By the early 1900s, men's suits were a woollen one-piece affair from the shoulders down to the knees. Women's suits were still pretty voluminous, consisting of an upper part with a short skirt covering baggy trousers down to the knees. Oppressive modesty was foremost and competitive swimming by women was really not expected. Until Annette Kellerman, that is.

Kellerman was a one-woman revolution. She was born in New South Wales in 1887 into a family of musicians. By the age of six, she had problems walking due to rickets and took up swimming to strengthen her legs.[81] By the age of 16, she was winning all the swimming races she entered, was swimming with fish in a tank to entertain crowds at the local aquarium and doing spectacular dives on stage in a play at the Melbourne Royal Theatre.

She travelled to England with her father to make her fortune, wowing the British by attempting to swim across the English Channel. Although she tried to cross it three times, she never managed it. Instead, she made her name by swimming seven miles down the muddy Thames during a period when pollution meant very little survived in the water.

Of course, Kellerman couldn't compete or perform in the swimsuits women were expected to wear; the weights in the hems would have made it suicidal. Instead, she made her own costume from a man's one-piece suit. Made of black knitted wool, this went from the shoulders down to mid-thigh, covering the tops of her arms as it would on a man; this was rather risqué as it showed off both her ankles and her knees. When she was asked to do an exhibition for the Royal family, she was told she'd have to cover up her bare legs, despite her protestations that in Australia it was completely normal to wear such a suit. Her solution was perhaps not what they expected. She sewed on a pair of tights to make a full-length one-piece suit. This did to the letter what they'd wanted but, unwittingly, it seemed to hug her every curve, enhancing her svelte, athletic body much like modern swimsuits do today. It was both modest and revealing at the same time. The swimsuit became her trademark.

She went to America in 1907 and began to perform in theme parks, showing off her swimming, diving and curves to packed audiences. But then she walked down Revere Beach off the coast of Boston in a short maillot-style swimsuit, intending to swim three miles out to the lighthouse. She bared both her arms and legs and heavily clothed women gathered around to point and gasp at her. A policeman was called and she was arrested for indecency.

In the court hearing that followed, Annette explained to the judge that she couldn't possibly swim in more clothes than she could hang on a washing line; it simply wasn't sensible. Swimming was the best exercise in the world, she said: after all, it had cured her of rickets. The judge was convinced. She could wear the short swimsuit as long as decency was maintained across the beach and she wore a robe right up until entering the water.

Hollywood noticed Kellerman and she starred in a $1 million silent movie called *Neptune's Daughter*. At about $25 million in today's money, it was the most expensive film ever made and the first in which the leading actress appeared nude. Kellerman used her fame to push forward the claim for women to be treated as equals to men and lectured across America on women's health, empowerment and swimming.

Figure 23. Annette Kellerman in her trademark full-length swimsuit circa 1905–1910.

It's possible that the Revere Beach incident was deliberately staged, but Kellerman had made her point – if you wanted to swim, you needed a swimsuit that clung to the body and reduced drag from the water. Women didn't swim at the first four Olympic Games but were finally admitted to the Stockholm Olympics in 1912. All wore a short one-piece swimsuit similar to Kellerman's but made from silk rather than wool, which could weigh as much as five kilograms when wet. The problem with silk was that, when it was wet, it became translucent. This meant that briefs also had to be worn underneath for modesty and fearsome chaperones followed the women wherever they went.

Back in Kellerman's home town of Sydney, a Scot named Alexander MacRae had set up a factory to make underwear, calling it first MacRae Hosiery Manufacturers and then MacRae Knitting Mills. His chief customer was the army and they needed a plentiful supply of woollen socks for their forces fighting abroad in the European war. After the war, he diversified into swimwear. He took the Kellerman-style swimsuit and slowly began to decrease

Figure 24. The victorious English 400-metre women's relay swimming team at the Stockholm Olympic Games, 1912 wearing silk swimsuits: Belle Moore, Jennie Fletcher, Annie Speirs, Irene Steer.

the amount of material. He pulled the shoulder straps inwards between the shoulder blades and fixed them together to stop them falling down while swimming. This also meant that the arms and shoulders were now unencumbered and free to propel the swimmer through the water. MacRae called his new swimsuit the 'Racerback'.

He became dissatisfied with his brand name: 'MacRae Knitting Mills' didn't really match the image of sleek swimmers cutting through the water. He held a contest with his staff to come up with a new name, offering a prize of £5 (about £300 today) for a new catchphrase. The winner came up with the following:

'Speed on in your Speedos.'

I hadn't realised that Speedo was Australian, but it makes sense if you say the slogan with an Australian accent and remember that Australians love to put an extra 'O' on the end of names (we once had an enthusiastic researcher from Queensland called Jon – everyone called him 'Jonno'). MacRae changed the company name to Speedo Knitting Mills.

The Racerback sold well for Speedo. The cut became increasingly severe across the shoulders and pushed acceptability a little too far. There was nearly a rerun of Kellerman's American experience for compatriot Claire Dennis at the 1932 Los Angeles Olympics when she was almost disqualified for wearing the revealing silk Racerback. After winning her heat in the 200 metres breaststroke in a record time of 3 minutes 8.2 seconds, a rival team complained that her suit showed 'too much shoulder blade'.[82] Protracted negotiations ensued and eventually she was allowed to continue. She won gold and knocked almost another two seconds off her time. Less was becoming more in swimming.

The troubled genius

In April 1896, just as Greece was packing away the first modern Olympic Games, Mary Evalina Carothers was giving birth to

Figure 25. Male swimmers competing in the Speedo Racerback swimsuit. Late 1920s.

the first of her four children in Burlington, Iowa. She called him Wallace and he grew up to be an intelligent boy fascinated by science. He was especially intrigued by the new world of the atom. His father, seemingly oblivious to his son's ambitions or talents, sent him off to learn shorthand. But once away from his father, Carothers started to do chemistry experiments in the basement of his college. His visceral love for chemistry took him to Harvard where a company called DuPont came looking for him.

Although the secret to the vulcanisation of rubber had been discovered 50 years earlier, the underlying science was still not well understood. People had given the type of material a name: 'polymer' from the Greek *poly* meaning 'many' and *mer* meaning 'parts'. There were two competing ideas on how rubber and polymers worked. The first, and most popular one, was that they were a large jumble of similar molecules in a big soup held together in a way yet to be identified.

The second rather crazy idea had been presented by Hermann Staudinger to his colleagues in Düsseldorf in 1926. It caused a sensation. He suggested that polymers consisted of enormous chains of molecules of many repeating parts. The delegates were astonished. One said: "We were as shocked as zoologists might be if they were told that somewhere in Africa an elephant was found who was 1,500 feet long and 300 feet high."[83]

Rubber-covered yarns had been created for textile use, but they were heavy and liable to snap easily. DuPont realised that there was value in research – if only they could understand the underlying nature of polymers, they could create a whole new profitable world of materials. The first prize they wanted was that of synthetic rubber. Chemistry would take up half their budget and Carothers was the man they wanted to lead the research team. But there was a problem: Carothers was reticent and admitted openly that he suffered from severe depression. Undeterred, DuPont persevered and got their man.

Carothers subscribed to the crazy Staudinger model of polymers and set out to prove the theory right. By 1930, his deep understanding of polymer science paid dividends. In a similar way to Goodyear's accidental discovery of vulcanisation, one of Carothers' assistants left a test tube of liquid polymer sitting on a shelf for a week. When it was found, Carothers realised that a new material had been created. It was the artificial rubber that everyone had been looking for with as many uses as Charles Goodyear had thought of: fuel pipes, shoe soles, wetsuits, exercise mats. Carothers called it 'polychloroprene'; DuPont called it 'neoprene'.

Carothers directed his assistants to work immediately on other polymer compounds and they struck lucky. Dipping a glass rod into a polymer mix drew out long fibres of a clear material, a bit like putting your spoon into hot milk and drawing off the skin. DuPont named it 'nylon' and sold it as a yarn to replace Japanese silk used in stockings; they quickly became known as nylons.

Sadly, Carothers' bouts of depression worsened and he never saw the success of the polymers he created: he killed himself with a cyanide pill in 1937.

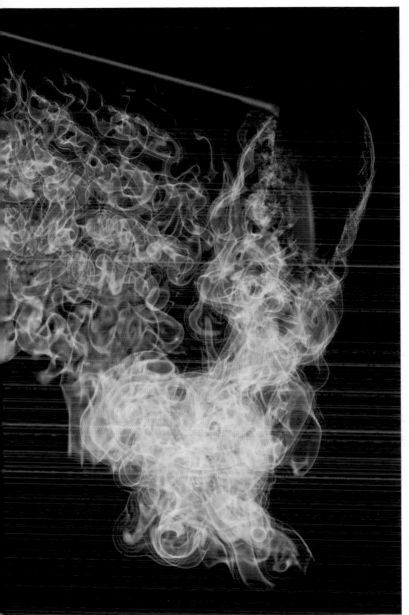

After oversize clubs were developed, the drag of the club head as it swung through the air became much more important. The image shows computational fluid dynamic analysis of the air flow around a PING G15 golf driver. This work led to the introduction of 'turbulators' on the top of subsequent PING driver crowns which significantly reduced drag, increased clubhead swing speed, and extended the driving distance of the ball.
© *John Hart.*

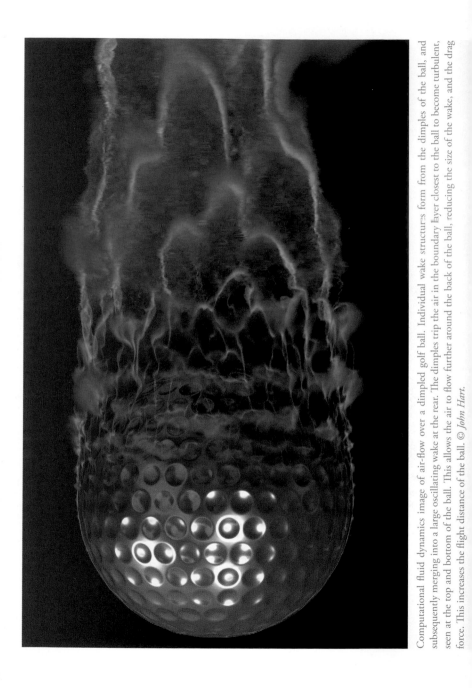

Computational fluid dynamics image of air-flow over a dimpled golf ball. Individual wake structures form from the dimples of the ball, and subsequently merging into a large oscillating wake at the rear. The dimples trip the air in the boundary layer closest to the ball to become turbulent, seen at the top and bottom of the ball. This allows the air to flow further around the back of the ball, reducing the size of the wake, and the drag force. This increases the flight distance of the ball. © *John Hart.*

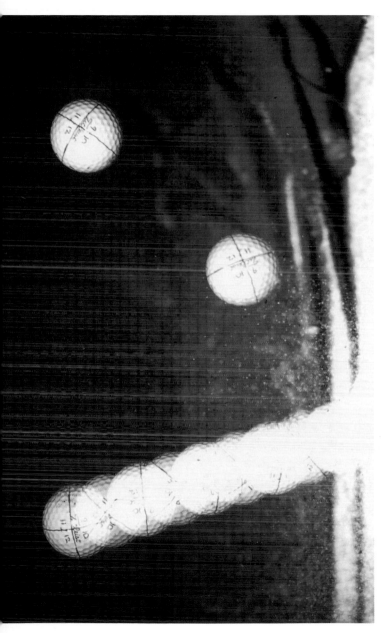

Stroboscope image of a golf ball hitting a golf green from a simulated shot from a driver; the images are five thousandths of a second apart. The ball enters from the right at 85 kilometres per hour (53 miles per hour) and rotates clockwise 33 degrees between consecutive images of the ball. This is equivalent to a backspin of about 19 revolutions per second (1,117 revolutions per minute). The ball rebounds to the left at around 14 kilometres per hour (9 miles per hour) so that its coefficient of restitution is 0.16. The backspin has been converted to topspin by the impact and the ball now rotates anti-clockwise at about 26 revolutions per second (1,550 revolutions per minute). If the turf conditions are right, the ball can retain backspin after impact and comes to a dead halt on the next bounce, even screwing back along the green. © Steve Haake

'Finite element analysis' model of a tennis racket. The model assigns a part to each component, splitting them into thousands of small elements each with a particular stiffness and material characteristic. This model, developed by Tom Allen for Prince Racquets, included the deflection of the strings, the internally pressurised ball, and the frame, something that had never been done before. © *Tom Allen, edited by John Hart.*

Airflow over a two-woman bobsled. The air flows over the nose of the sled, strikes the driver's helmet and then swirls around the driver and brake woman. The areas in red show where the air is moving fastest; the light spot on the front of the driver's visor indicates where the air first hits her, before flowing around the sides. © *John Hart*.

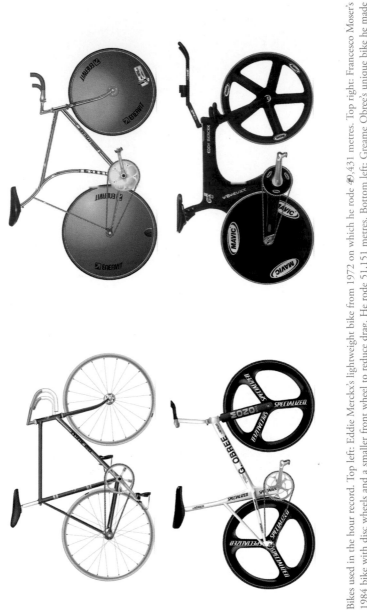

Bikes used in the hour record. Top left: Eddie Merckx's lightweight bike from 1972 on which he rode 49,431 metres. Top right: Francesco Moser's 1984 bike with disc wheels and a smaller front wheel to reduce drag. He rode 51,151 metres. Bottom left: Greame Obree's unique bike he made himself in 1993 with unusual 'T-bar handlebars and on which he rode 52,713 metres. Bottom right: Chrs Boardman's 1996 bike on which he set an hour record in Manchester of 56,375 metres. Illustrations © *James McLean.*

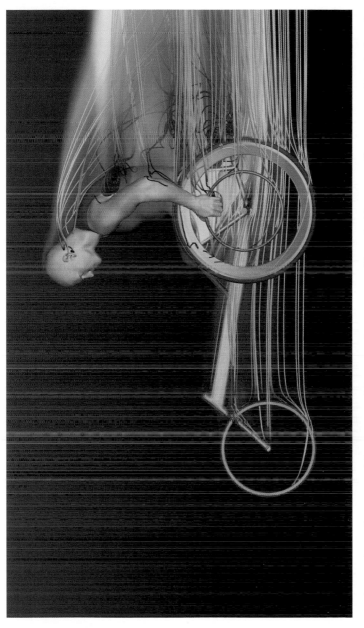

Aerodynamic drag makes up around three quarters of the resistive forces experienced by a wheelchair racer; rolling resistance of the wheels on the ground makes up most of what's left with a small proportion due to inefficiencies in the bearings and the frame. © *Tom Hart*.

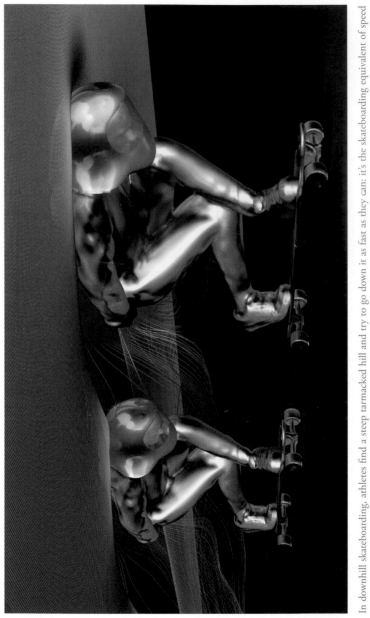

In downhill skateboarding, athletes find a steep tarmacked hill and try to go down it as fast as they can: it's the skateboarding equivalent of speed skiing. Skaters achieve speeds in excess of 120 kilometres per hour by adopting an aerodynamic tuck position and drafting each other. The picture shows two drafting skateboarders wearing aero-helmets. The flowlines show that the air flows smoothly over the skaters' helmets but swirls chaotically behind their legs. © *John Hart.*

Less is more

Carothers' laboratory continued on without him with a strategy to create new yarns and to improve the ones they already had. A more versatile synthetic rubber was still the target. While neoprene was great, it came out as a flat sheet which tended to limit its use. Nylon had become ubiquitous but it still didn't have the flexibility of rubber. It took another 20 years until, in 1958, a DuPont scientist called Joseph Shivers finally created what they'd been searching for all this time. It had a name more like a breakfast cereal: Fibre K. It could stretch to over five times its original length without breaking, it was strong, light and relatively immune to chlorine or oils.[84] The name Fibre K didn't catch on: instead, they called it 'Lycra'.

Lycra spawned a whole new world of possibilities for clothing. I have early memories from the mid-1970s of my underwear and trousers scratching me around the waist. I'd investigate the waistband to find tiny ends of broken grey rubber strands and I'd pull at them to get them out, only to make things worse. I can also remember the embarrassment of jumping into a swimming pool only to find my swimming trunks floating off towards the deep end. Lycra was my saviour and the saviour of thousands of skinny kids like me: not only did it replace the fraying rubber, it also helped me keep my pants on.

In the 50 or so years between Annette Kellerman and my own ill-fated attempts at diving, swimsuits had dramatically reduced in size. During the Second World War in America, the clothes industry had been mandated by the government to conserve material. This was just the excuse the designers needed. Men's suits shrank so that their chests were bared down to the navel, considered by some the line you couldn't cross. By the 1948 London Olympics, this was the norm. Women's suits were reduced in size as far as they dare; on the beach, the one-piece became a two-piece with a bare midriff. This shrank further after the war to become the bikini, named after the Bikini Atoll where atomic bomb tests were carried out. The bikini, it was said, had an explosive effect on the viewer.

The bikini was hardly something to swim in competitively, however. Speedo's performance swimwear reduced in size until the term 'Speedos' became synonymous with the tiny pants used by men from the sixties onwards. My Australian friends call them 'budgie smugglers'. Woollen swimwear was baggy, soaked up water and increased drag. Better to reduce coverage as much as possible down to bare skin which had the added advantage of being naturally waterproof. Women's suits had minimal straps, were cut as low as possible on the chest and higher and higher over the thighs. Peak thigh occurred at the 1984 Olympics in Los Angeles with the US women's team looking more like Baywatch actors than competitors.

The turning point came when Speedo introduced nylon and Lycra to their swimwear. This combination of materials gave athletes suits that were light, tight-fitting and low in water absorbency. By the 1992 Olympics in Barcelona, Speedo launched their new S2000 suit for women with the claim that it reduced drag by a massive 15 per cent.

The stretch of Lycra – with the generic name 'elastane' – combined with close-cut design kept the clothing tight to the body just as Kellerman had wanted way back in the early 1900s. The thinking about body coverage went into reverse: the back was filled in and the chest and neck covered up to stop air pockets. The men were not quite convinced and still preferred the minimalist look, but a Speedo research paper written in 1997 presaged what was about to happen: 'If a swimmer asked us to make a catsuit, we would do it. However, psychologically this is quite a big jump… Maybe this will be done in the future.'[85]

The future arrived quickly at the Sydney Olympics just three years later. World record holder Pieter van den Hoogenband of the Netherlands wore a suit that went from the waist all the way down to the knees. American Josh Davies went further and wore something similar to the women, covering himself from the neck to the ankles but keeping the arms and shoulders free. Ian Thorpe went the whole way and covered his body, arms and legs. Only his hands, feet and head were left uncovered.

Speedo's new swimsuit for the Games took sharkskin for its inspiration – at least according to the marketing. They claimed the suit reduced drag even more than the S2000. The drag on a swimmer is similar in nature to the drag on a golf ball or a bobsleigh but is usually split into three components. Firstly, there is pressure drag: this is where the water moves out from the front of the swimmer, round the sides and to the back, leaving a large wake behind. Secondly, there is skin friction, caused by energy losses in the boundary layer directly in contact with the swimsuit. Lastly, there is wave drag, produced when the swimmer creates a little bow wave at the front of the body, much like a boat.

Speedo tried to mimic the sharkskin grooves with a beautifully complex weave. A real shark's skin is covered with small scales between a tenth and a half a millimetre long in repeated patterns across its body. Called 'denticles', meaning 'little teeth', they have grooves that align water flow along the shark's body from front to back and limit sideways eddies that would waste energy during swimming. They also lift the boundary layer away from the surface up onto the peaks of the grooves, a bit like someone lying on a bed of nails. This limits the area of skin contacting the fast-moving water to mere points and reduces the skin friction dramatically.

Speedo claimed that the weave of their material did all this and subsequently reduced drag by over seven per cent. They called their new innovation 'Fastskin'. It certainly made an impression with the swimmers and over 80 per cent of medallists at Sydney used the suit.[86] Speedo's Fastskin II, developed over the next four years for Athens in 2004, claimed to reduce drag by a further four per cent through better fit. Speedo's Aqualab were at their creative best during this time and worked hard on a new innovation for the next Olympics in Beijing. They didn't disappoint.

Speedo used a laminate of polyester weave covered in a polyurethane finish. Polyurethane is smooth, stiff and hydrophobic, meaning that it repels water. Some plants' leaves work in the same way. I can often see the hydrophobic effect on the large hand-like leaves of the lupins in my garden. Water is repelled from the leaf surface and rolls down to the centre to form a bubble, leaving the rest dry. The bubble is nature's way of keeping the contact area between the water and the surface to a minimum: if you see a random bubble on a surface, rather than a flat puddle, then it tells you that it is hydrophobic.

Speedo sewed panels of their stiff hydrophobic laminate down the sides of their nylon-elastane suits, calling it the LZR – pronounced 'laser'. The suit compressed the enclosed body to reduce pressure drag and repelled water to reduce skin friction.

"I feel like a rocket," said Michael Phelps.[87]

Figure 26. Launch of the LZR swimsuit in New York, February 2008.

Athletes clamoured for the suit and rival brands reluctantly allowed their sponsored athletes to wear it. More than 90 per cent of gold medals at Beijing were won by athletes wearing the LZR. Phelps won eight on his own. Rival company Arena called the suits tantamount to technical doping but immediately set about developing their own, as did adidas, Nike, TYR and Mizuno.

Phelps vs Biedermann

In 2009, Paul Biedermann started to wear a new Arena suit called the X-Glide. If the polyurethane panels were that good at reducing drag, went the logic, then why not make the whole suit out of it?

The atmosphere prior to the World Championships was feverish and world records had been hinted at in training. But the critics were becoming more vocal. Rebecca Adlington, winner of two gold medals in Beijing wearing a LZR, refused to go the next step and wear a full polyurethane suit. The national federations began to ban the suits in their own countries and, despairing of the FINA executive, voted at FINA's own AGM on the eve of the Championships to ban the swimsuits once the tournament was over.

But it was too late for the Championships themselves. A study by Enrique Nueva showed that half chose the full polyurethane suits made by Jaked and Arena.[88] Between them, the two companies won 80 per cent of the medals. The swimmers astonished the world by setting world records in the heats, only to beat them again in the finals. A swimmer told me that FINA paid $10,000 for each world record lasting longer than a day: they must have feared bankruptcy.

Biedermann recognised the benefits the polyurethane suit gave him and said it improved each 50-metre lap by about three tenths of a second; this would give him an advantage of 1.2 seconds in a 200-metre race. Biedermann beat Phelps – wearing his LZR – by exactly this amount.

Phelps' controversial coach Bob Bowman asked angrily how Biedermann had managed to knock off a full four seconds in less than a year when it had taken Phelps five years of hard training to do less. The Biedermann–Phelps story became the focal point of the whole swimsuit debate: new technology against old technology; swimming legend against new kid on the starting block.

"It'll be great to swim against him without the swimsuit," said Phelps.

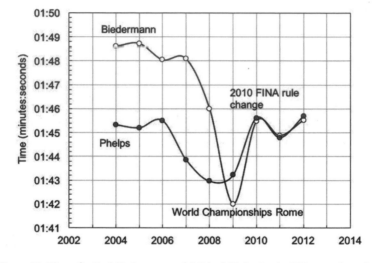

Figure 27. Times for Paul Biedermann and Michael Phelps in the 200-metre freestyle. Phelps wore Speedo's Fastskin from 2007 while Biedermann wore Arena's X-Glide from 2009. *Data courtesy of Leon Foster.*

Bowman was both right and wrong about Biedermann; if he'd looked at the stats he'd have seen that Biedermann had already been on a massive trajectory of improvement. Between 2007 and 2008, he'd dropped a whopping two seconds despite wearing the inferior legging-style swimming shorts made by adidas. Of course, when he'd taken up the X-Glide in 2009, his time had dropped by a further four seconds. Phelps, on the other hand, had progressed from shorts to the Fastskin full bodysuit a year earlier than

Biedermann at the World Championships in 2007. Accordingly, his time had dropped by 1.6 seconds and he'd set a new world record. This continued into the Olympics in Beijing with another one second improvement and another world record. The point is that Phelps and Biedermann had similar experiences with the suits, it's just that Phelps had it first, but then decided to be loyal to the LZR rather than go for the full polyurethane suit.

When the two met in Rome in 2009, Biedermann had just acquired the X-Glide and was on fire; Phelps had taken six months off. The inevitable happened and Biedermann stormed to victory.

Once the suits had been banned by FINA in 2010, the men reverted to 'jammers', a shorts-style swimsuit that went only from the waist to the knee. How did the two get on without the advanced swimsuits? For the next two years, their best times were separated by mere tenths of a second. The result of Phelps' desire to race Biedermann without the full body swimsuits was this: they were almost identical.

How did the suits work?

The data on whether the suits worked is unequivocal: performances for both men and women in most events increased when the full bodysuits came out. There were dramatic increases in swimming speed in the freestyle for the period between 2007 and 2009 when the full body polyurethane swimsuits were used. The men saw bigger increases than women, probably because they covered up more of their previously bared flesh, and the women were already half covered anyway. Interestingly, the improvements lessened with distance so that swimmers in the 800 and 1,500 metres saw almost no change at all. Sometimes the swimsuits worked and sometimes they didn't.

Even after the ban, researchers carried on looking into the suits, desperate to understand if and how they actually worked.

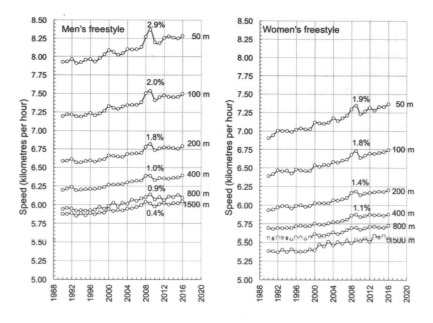

Figure 28. Swimming speed in the men's and women's freestyle from 1990, showing the average (mean) of the top 25 swimmer performances in each year. The increase in speed between 2007 and 2009 is shown for each distance. *Data courtesy of Leon Foster.*

Early research into swimming had concentrated on the fact that a swimmer is more of a wobbly mass than a sleek torpedo. In the 1970s, Yuri Aleyev used a cable to pull a selection of naked women through the water to look at the movement of their skin as water flowed over it.[89] He found that skin folds rippled down the body and noted that this increased skin friction through the water. Covering up the skin would reduce this effect.

Huub Toussaint of Amsterdam suited up 13 men and women in different swimsuits and found a small statistically insignificant improvement of two per cent in the Fastskin suit.[90] The consensus of many researchers was that the main effect of the suits was to tuck the body in and reduce the cross section of the body flowing through the water. Brian Dean and Bharat Bhushan from the fantastically named Nanoprobe Laboratory for Bio- and Nanotechnology and Biomimetics at Ohio State University were certainly sceptical about the ability of textiles to mimic sharkskin. They gave a detailed review of the hydrodynamics of sharkskin and Speedo's claims, concluding that, "given the compromises of riblet geometry made during manufacturing, it is hard to believe the full extent of the drag reduction".[91]

Given the compression that the suits gave to the human body, Kainuma[92] suggested that the suits were so tight that blood circulation was impeded. This might explain why performance vanished for the longer-distance swimmers who would spend much longer swimming in the suits. Sprint swimmers rely on anaerobic respiration, getting a large proportion of their energy supply directly from the muscles. Longer-distance swimmers, on the other hand, get more of their energy from aerobic respiration so, while drag might be improved, their energy efficiency might be worsened. This would cancel out the reduced drag benefits of the suits.

Given the hydrophobic nature of polyurethane and its ability to reduce skin friction, my suspicion is that the glide phase is where most of the gains would be made. To test this hypothesis, I asked Biedermann to do some experiments in the pool in Dresden. I asked him to dive in and glide as far as he could without swimming, first in a simple pair of trunks similar to the ones that Mark Spitz wore in the 1970s (complete with stars and stripes motif), then in his X-Glide and finally in a pair of approved post-2010 jammers. He repeated the test three times. In the trunks, he managed to glide just under 20 metres before coming to a stop. He went almost 25 metres in his X-Glide, almost half the length of the pool. The new 2010-approved swimming shorts were bang in the middle with a glide of just over 22 metres.

Peace after the swimsuit wars

People tend to forget that Paul also went on to break the world record in the 400-metre freestyle at the Championships in 2009, shaving just a hundredth of a second off Ian Thorpe's time set in Manchester in 2002. He was honest enough to admit that the suit gave him at least two seconds' advantage over his old one.

"I think the suits destroy a little bit of the real sport," he said at the time. "It's not any more about technique; it's not any more about good starts or turns; it is just, put this thing on and feel really, really fast in the water. I really believe all the new suits should be banned."[93]

The suits were banned by FINA from 2010 and the swimsuit wars were over. Although the full body swimsuits improved performance significantly, especially in the shorter events, I think FINA got it wrong to ban them: there would have been a step change but then performances would have levelled off again. The 2009 peaks had some performances that would take years to beat. In 2011, I used performance data to predict when the 2009 records set before the rule change would be broken with the downgraded suits. Paul Biedermann's and Federica Pellegrini's 200-metre records, for instance, might take until 2017/18 to be broken.[94] This is still true as I write this.

	Men		
Distance	2009 Record	Predicted date of new world record following 2010 ban on suits	Outcome (✓Prediction correct; ✗prediction wrong)
50 m	César Cielo (BRA) 18 December 2009 20.91 s	Between 2019 and 2023	✓ Still not broken
100 m	César Cielo (BRA) 30 July 2009 46.91 s	Between 2010 and 2017	✗ World record due
200 m	Paul Biedermann (GER) 28 July 2009 1 min 42 s	Between 2017 and 2018	✓ Still not broken
400 m	Paul Biedermann (GER) 26 July 2009 3 min 40.07 s	Between 2013 and 2019	✓ Still not broken
800 m	Zhang Lin (CHI) 29 July 2009 7 min 32.12 s	Between 2011 and 2026	✓ Still not broken
1500 m	Grant Hackett (AUS) 29 July 2001 14 min 34.06	2012	✓ Broken Sun Yang (CHI) 4 August 2012 14 min 31.02 s

Women			
Distance	2009 Record	Predicted date of new world record following 2010 ban on suits	Outcome (✓ Prediction correct; ✗ prediction wrong)
50 m	Britta Steffen (GER) 2 August 2009 23.73 s	Between 2012 and 2017	✓ Broken Sarah Sjöström (SWE) 29 July 2017 23.67 s
100 m	Britta Steffen (GER) 31 July 2009 52.07 s	Between 2013 and 2021	✓ Broken Cate Campbell 2 July 2016 52.06 s
200 m	Federica Pellegrini (ITA) 29 July 2009 1 min 52.98 s	Between 2010 and 2017	✗ World record due
400 m	Federica Pellegrini (ITA) 26 July 2009 3 min 59.15 s	Between 2010 and 2014	✓ Broken Katie Ledecky (USA) 9 August 2014 3 min 58.86 s
800 m	Rebecca Adlington (GBR) 16 August 2008 8 min 14.10 s	2011	✗ Broken late Katie Ledecky (USA) 3 August 2013 8 min 13.86 s
1500 m	Kate Ziegler (USA) 17 July 2007 15 min 42.54 s	Between 2010 and 2012	✗ Broken late Katie Ledecky (USA) 30 July 2013 15 min 36.53 s

World records after the 2009 World Championships in Rome in the freestyle where 43 records were broken using the new swimsuits. After FINA banned the swimsuits in 2010, the data showed that it would take years for the world records to be broken again. Ticks show where the prediction was right, crosses where the prediction was wrong. The men's 100 metres and women's 200 metres are due to be broken any time now.

Out of the 12 freestyle events I analysed, my prediction was right in eight of them. Two records in the women's 800 and 1,500 metres should have been broken by 2012 but weren't broken until 2013 when American Katie Ledecky burst onto the scene. There are two records that are certainly overdue: the men's 100 metres and the women's 200 metres. Keep an eye out, they should be broken very soon. The swimsuits did make a difference, but either FINA shouldn't have banned them or banned them and reset the world records to the 2007 value.

The example of swimming shows how materials and design can come together to make a big difference when it comes to improving performance. The impact was to reduce the energy-sapping drag of the swimsuit, effectively making the event easier.

The sport in the next chapter underwent a similar transformation in the 1990s and sought not only to decrease the energy losses but to optimise the energy input of the athlete. The outcome was dramatic.

NINE

Framing the problem

27 OCTOBER 2000: The Manchester Velodrome, 5 p.m. It had been a wet and windy day, pretty much par for the course. The weather forecast showed a deep depression crossing Scotland pulling a jagged rain front along with it like a broom to sodden the towns and cities to the south. The silver lining in the rain clouds was that the air pressure would be low. This meant the air density would also be low: perfect conditions for a bike race.

The atmosphere inside the velodrome was electric. It was surprisingly warm and the damp coats of the 2,000 excited spectators raised the humidity somewhat, reducing the air density just a little bit more.[95] The crowd erupted as the gladiator they'd come to see rode out of the entrance tunnel: Chris Boardman MBE. Here was a true British hero, world record holder and winner of Olympic gold in the four-kilometre pursuit at the Barcelona Olympics. He was the arch rival – so the press would have you believe – of the maverick Graeme Obree. Chris was here for his swansong with an attempt to beat the great Eddy Merckx's one-hour record of 1972. It was simple stuff: cycle as fast as you can for an hour. If he could manage 198 laps and go further than 49,431 metres, then everlasting glory would be his. At least until the next attempt.

The velodrome quietened with only the occasional cough breaking the breathy silence. Then with a roar he was off, quickly reaching one kilometre after just 78 seconds, only seven seconds down on Merckx's start. After a quarter of an hour, the large black digital scoreboard predicted a win by 350 metres. The crowd went wild. Peter Keen, Boardman's coach and key conspirator, stalked the track's edge, holding up his fingers to indicate how fast or slow the pace was compared to their plan. His position on the straight let Boardman know how well he was doing overall: if he was upstream of the middle, Boardman was ahead of Merckx; if he was downstream, he was behind.

As the half hour passed, those who understood these signals knew that Boardman was well ahead. But as the laps ground on, Keen gradually shuffled back downstream until, with eight minutes to go, he passed the middle. Boardman was behind target.

The commentator whipped up the crowd, informing them that Boardman was two seconds behind, now three point four seconds behind and helpfully screamed that 'Time is running out!' Like the penultimate scene of any good thriller, our hero was hanging off a cliff, his fingers slipping slowly one by one off the edge. The crowd screamed as if noise alone could propel Boardman along. With only four minutes to go, Keen had walked so far back down the straight that he was almost at the bend and Boardman was well behind Merckx. Boardman's wife, Sally, could take it no longer and ran down to the trackside to scream her support.

But then Keen stopped. On the next lap, he moved slowly back towards the finish line. And again on the next one. As our hero pulled himself up from the cliff edge, the anguish on Sally's face turned to hope. Boardman seemed to grow in strength and wound up for a last agonising push. The gun went on the hour with a double crack as the board showed the result: 49,441 metres, just ten metres ahead of Merckx's record.

The crowd went crazy, Sally Boardman and Peter Keen hugged in tears and Chris Boardman circled gently around the track in triumph.

Find the video and watch it: I defy you not to cry.

Emotional as it was, the weird thing is that Boardman had

already beaten the record. Not just beaten it, but absolutely smashed it. Just four years previously, Boardman had cycled a massive 56,375 metres on exactly the same track. That's almost 28 laps more. Hadn't anyone noticed? Why was he so slow today? Was there something wrong with his bike? Bad bearings? Flat tyres? Wrong gearing? Just what was going on?

Surely, all he needed was a better bike.

Playing it safe

Almost everyone remembers their first bike. I have an early recollection of something with fat white wheels when I was around eight, but my first real bike was when I was 12: a yellow, steel-framed Claud Butler with drop handlebars and five derailleur gears operated by a single lever on the down tube. I hammered it around the village and was forever fixing punctures and taking it to the bike shop in town to have the wheels re-trued. My bike was my liberty, my first bid for freedom.

The development of the bicycle had a stuttering start 200 years ago. Karl von Drais, from Karlsruhe in Germany, is usually given the honour of having created the first bicycle-like machine. In the early 1800s, he'd played around with four-wheeled carriages that used human muscle and complex levers to propel themselves along. It was clearly impractical and his peers found it hilarious. Then he had an epiphany: he ditched the whole lot and went back to just two wheels. He put a simple wooden beam between them, put a cushion on it to sit on and attached a T-bar steering mechanism. The rider propelled himself with his feet (men only of course) so that he could cruise along at eight or nine kilometres per hour, perhaps double walking speed.

His friends were more impressed this time. Actually, they were astonished. How did it stay upright? Why didn't it just fall over? Where Drais got the simple idea to try two wheels has never been adequately explained, but he had stumbled upon the key to a bicycle's stability – its steering. This is the first thing you find out when you learn to ride. If you fall to one side, your first instinct is to turn your steering wheel away from it. All this does, however, is make you fall faster. The trick is to turn into the fall: this brings the wheels back under your centre of mass and keeps you upright.

In 1818, Drais called his human-powered machine a 'velocipede' from the Latin for 'swift of foot' (the French would later shorten this to *vélo*). Those who tried it were variously excited, frustrated and embarrassed – by hills, poor roads and magistrates who fined anyone caught riding on the pavement. Enthusiasm dissipated. The world wasn't quite ready for the bicycle and this embryonic version flashed briefly before vanishing again for 40 years.

Figure 29. Drawing of an early Karl von Drais velocipede.

The odd enthusiast still made velocipedes, of course, with some even being exhibited at the Great Exhibition in 1851. Around 1867, the Olivier brothers commissioned Paris blacksmith Pierre Micheaux to attach cranks and pedals directly to a velocipede's front wheel. They redesigned the whole machine to give it a horizontal handlebar attached to front forks like those today, an adjustable saddle on a leaf spring and even a rudimentary brake. The pedals were a revelation, allowing the rider to cruise along at ten kilometres per hour or more. It was as fast as a horse and didn't need feeding.

The addition of pedals drove a second boom in cycling as other manufacturers adopted the design. The heavy wooden struts and beams were replaced by a single wrought iron beam from the top of the forks to the rear axle, with the seat on a thin horizontal metal beam. This triangle offered a glimpse of the diamond frame that would be used by the likes of Merckx and Boardman a century later.

The craze spread across the world and hit America where sales of thousands created work for blacksmiths, salesmen and lawyers filing patents and suing for breaches. By 1869 the wheel had been reinvented – this time with fine wire spokes and a groove in the rim to accept a strip of rubber. Races started to appear, such as the 80-miler from Paris to Rouen, won in a time of ten and a half hours by James Moore of Suffolk. Moore – an adopted Parisian – had used newly invented ball bearings in the axle of his front wheel. Previous axles had been a simple sheath of brass surrounding the bar of the crank running through it, but it quickly wore out to create high friction and a wobbly wheel. With his new bearings, Moore won the race by 15 minutes.

Initially, the machines were obscenely heavy, but this soon decreased when the wrought iron frame was replaced by hollow steel tubing. Designers quickly realised that because of the crank's direct drive, the bigger the wheel the further the bike would go for one turn of the pedals. The front wheel grew in size with the only limit being the length of the rider's inside leg.

By the early 1880s, the bicycles had huge front wheels over a metre in diameter and had been renamed high-wheelers; today we know them as penny-farthings because of the disparity between the enormous front wheel (the penny) and the tiny rear wheel (the farthing). One bicycle that stood out for its craftsmanship was the Ariel, made in Coventry by James Starley and William Hillman. It had a single metal hollow down tube for its frame and radially spoked wheels. It had a rear brake activated by a twist of the handlebar, adjustable pedals and a sprung saddle.

Figure 30. Penny farthing racers in 1888 where riders matched the direct-drive front wheels to their leg length. The bigger the wheel, the further the distance travelled for one rotation of the crank.

These penny-farthings were pretty exclusive, not least due to their cost. At £8 – about £800 today – they were hardly the transport of choice for the common man and certainly not for the common woman. The high-wheeler was scary to ride – if the front wheel hit an obstacle, then the high centre of mass automatically tipped the rider head first towards the ground from over two metres. Riders proudly called this 'taking a header'. Injuries and even fatalities were common.

The cycling business was in danger of stagnating again, but was about to get a stimulus to keep it going. What James Starley didn't realise was that he would be one of the key players in the next

THE ROVER SAFETY
BICYCLE (PATENTED)

Safer than any Tricycle, faster and easier than any Bicycle ever made.
Fitted with handles to turn for convenience in storing or shipping. Far and
away the best hill-climber in the market.

Mᴬᴺᵁꜰᴬᶜᵀᵁᴿᴇᴅ ʙʏ

STARLEY & SUTTON,
METEOR WORKS, WEST ORCHARD. COVENTRY. ENGLAND.

Figure 31. Left: early advert for the Rover Safety bicycle, circa 1885. Right: John Boyd Dunlop riding a safety bicycle complete with his pneumatic tyres around 1915 when he was /5 years old.

big thing for cycling. He called his bicycle of 1885 the Rover. It had two spoked wheels of similar size, a diamond frame, a crank in the middle of the bike attached to a chain to drive the rear wheel, an adjustable saddle and handlebars attached to front forks. Importantly, the rider could start and stop with their feet on the ground and the chances of doing a header were slim.

It quickly became known as a 'safety' bicycle – used to distinguish it from the patently unsafe high-wheeler which was renamed by its devotees 'the ordinary'. The safety bicycle wasn't a real bicycle, they said; it would never catch on.

Starley probably had in mind as customers the same rich gentlemen of leisure who bought the high-wheeler, but the new Rover appealed to those who had been excluded because they weren't tall enough, were scared of falling or weren't men. Women loved the Rover and it gave them a sense of freedom for the first time. The bike forced fashions to change and women rejected the heavy skirts of the day for outrageous knee-length trousers they called bloomers.

The Rover was bought by rich and poor alike and it became the people's nag. The bicycle industry boomed and the Rover became the most popular bike across the world: the word for bicycle in Polish is still *rower* (pronounced rover).

The cycling boom

But there was just one last piece to the puzzle. Roads and turnpikes across Britain had deteriorated since the advent of the railways in the 1830s and cycling was a bone-shaking experience. John Boyd Dunlop realised that his son's tricycle could be made more comfortable by giving it pneumatic rather than solid tyres, so he glued two pieces of rubber together, attached them to his son's wheels and inflated them to almost balloon-like dimensions. This dampened the vibration from the road and his son cycled off in delight. But there was a second, unexpected advantage: the bike was much faster.

By 1890, Dunlop had gone into business and racers of the day immediately adopted the new technology. The 100-mile record was smashed in five hours, 27 minutes. Edouard Michelin improved the tyre further by making it detachable, leading to the ubiquitous 'clincher' tyre we use today. The inflatable part became an inner tube and punctures could now be repaired by the side of the road. The final piece had arrived – the bicycle was now safe, comfortable and fast. In fact, it was the fastest thing on the roads (it still is in many cities).

Races proliferated and Pierre de Coubertin was quick to embrace cycling for his first Olympic Games in 1896. Athletes used single-gear bikes that wouldn't look out of place today with hollow-tubed diamond frames, sloping front forks and drop handlebars. The Greeks built a brand-new velodrome for it in Piraeus, in the place where the Olympiakos football stadium stands today. The road race took the riders 87 kilometres from Athens to Marathon and back, ending at the new velodrome. The Greeks were passionate about marathon races, as they were born from the story of the messenger Philippides who ran from Marathon to Athens to announce the defeat of the Persians. It defined both glory and sacrifice, so when the Greek runner Spyros Louis had won the first Olympic marathon the week before, the Greeks were understandably ecstatic. With five of the starting line-up for the cycling road race being Greek, they were optimistic of another win.

At midday, an enthusiastic crowd cheered off the cyclists from the start line just on the edge of Athens; they raced away so rapidly "did they seem to fly over the ground".[96] The Greek Aristides Konstantinidis was first to reach the 40-kilometre post in Marathon after about 75 minutes. Following the rules of the day, he dismounted to scribble his name down on a parchment and remounted to race back to Athens. He was chased by two riders, Goedrich from Germany and Battell from Britain. Konstantinidis' bike broke and they raced past him. His coach caught up with him and they swapped bikes so that, with a supreme effort, Konstantinidis caught and overtook his rivals just outside Athens with only nine kilometres to go to the finish in Piraeus.

He flew through the centre of Athens to cheers from the thick crowds. Their cheers quickly turned to groans as he fell at a tight corner, cutting his arm and fatally damaging his second bike. But he wasn't done yet: as the British rider overtook him, a friend thrust a third bike at him.

Battell would surely have won, but not far from the stadium he crashed too, cutting himself badly. Konstantinidis slipped past and into the velodrome just as the Greek Royal family were settling into their seats.

"Mr Konstantinidis," said the race report, "covered with dust, begrimed and dirty, his whole appearance showing traces of his various accidents made a triumphant entrance, greeted by the enthusiastic cheers of the whole audience." His time was three hours and 22 minutes. Goedrich arrived 20 minutes later with Battell limping in behind him, battered and bruised.

Cycling was here to stay.

Figure 32. Léon Flameng (left) and Paul Masson (right), winners of six gold medals at the 1896 Olympic Games.

Cometh the hour, cometh the man[97]

Companies realised that, with the booming cycling industry, they could promote themselves by setting up races. Velodromes popped up in France, Britain and the United States and hosted races lasting up to six days, the longest they could go without breaking for the Sabbath. This allowed for a constant stream of crowds, eager to place wagers and spend money.

The supreme cycle race of all, however, was the Tour de France. Originally, it was no more than a marketing stunt to sell copies of the newspaper *L'Auto*, the ancestor of the modern French sports daily *L'Equipe*. Although *L'Auto* focused on the *Automobile* and *Cyclisme,* the paper's strapline showed that it covered most popular sports of the time: *Athlétisme, yachting, aerostation* (ballooning), *escrime* (fencing), *poids et haltères* (weights), *hippisme* (horse racing), *gymnastique* and *alpinisme* (mountaineering).

The paper's owner and editor, Henri Desgrange and Georges Lefèvre, had actually copied their principal rival, *Le Vélo*, who had already promoted long-distance races. Races with stages were best because they would need days of coverage for the pre-race build-up. The first Tour de France in 1903 was an overwhelming success and had 144 riders fighting it out over a gruelling 2,248 kilometres. Only 21 finished the six stages, with Maurice Garin the overwhelming winner in a time of 94 hours 33 minutes, three hours ahead of his nearest rival.

Desgrange saw the Tour de France as *his* race and he ran it tyrannically for three decades. The period saw many technological advances but Desgrange was reluctant to let them in, preferring to see his athletes suffer for their victory.

Desgrange saw another chance for personal glory – the hour record. The International Cycling Association had been formed and was to stage the first official hour record. Desgrange wanted to be the first in the record books – all he had to do was complete the hour. At the Buffalo Velodrome (named after the Buffalo Bill Wild West shows held there) and riding a safety bike, Desgrange managed 35,325 metres. The record lasted just four days before it was beaten.

The International Cycling Association didn't last long either as it was pushed aside in a coup by the Union Cycliste Internationale (UCI) in 1900. As a reward, one of their first responsibilities was to look after the hour record: it was something they'd probably come to regret.

The hour record continued to be broken and, by 1914, Oscar Egg had raised the distance to 44,247 metres. Fausto Coppi, the great Italian rider, raised it further in the 1940s and by the end of the 1960s it had reached 48,653 metres, set by Ole Ritter in Mexico City in 1968. The next high-profile rider to attempt it was the Belgian Eddy Merckx. In his first year as a professional, Merckx had been asked what his ambitions were. "To win the Tour de France," he said, "and set a new hour record."

Merckx's plan for the 1972 season had been to do fewer races than normal and then put in dedicated training for the hour record at its end. But tours and races begged him not to drop out and he did as many races as ever, winning 50. Like Ritter, Merckx opted for an open air velodrome in Mexico City for his attempt, in part for its high altitude and low air density. This would reduce drag by about 25 per cent. The downside was that there was less oxygen in the air which wasn't good for an endurance cyclist.

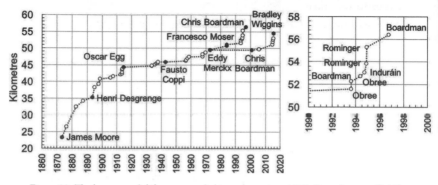

Figure 33. The hour record: kilometres cycled in an hour since 1873. Records were officially recorded from 1893 onwards when Henri Desgrange cycled 35,325 metres at the Buffalo Velodrome in Paris.

Merckx's solution was to train wearing a mask delivering lowered oxygen levels which would force his body to make more red blood cells to compensate. Merckx had a special bike made by Ernesto Colnago who lightened the bike by drilling out the chain and putting 48 holes in the handlebars. The stem was fashioned out of titanium, a tricky task in those days. He contemplated putting helium in the tyres to lighten it further but couldn't find any in Mexico. The resulting bike weighed a mere 5.5 kilograms.

Merckx had decided he wanted to do the race in style by breaking the 10- and 20-kilometre records at the same time. This had never been attempted during an hour record and would ensure that Merckx became the best rider of all time.

The weather was near perfect, warm enough to bring the air density down a little further and with minor gusts of wind of only a few miles per hour. Just before nine in the morning, Merckx set off at a blistering pace, just six seconds outside the world record for the first kilometre. At well over 50 kilometres per hour, he would easily beat Ritter's record. He beat the 10-kilometre record by five seconds and the 20-kilometre record by 11 seconds; only then did he drop his speed. It continued to fall, but he'd already done enough. He added three quarters of a kilometre to the hour record, pushing it to 49,431 metres. He'd beaten Ritter convincingly three times in one race.

"I will never try it again," he said afterwards. He never did.

Figure 34. Eddy Merckx' bike from 1972. *Illustration* © *James McLean.*

The mavericks

Some felt that Merckx could have gone much further had he not set off quite so fast. Even so, his record would last for another 12 years until it was beaten by Italian Francesco Moser in 1984. He and his team considered all the things they could do to improve performance and made 15 experimental bikes. The resulting design was stunning.

It had disc wheels with the front wheel smaller than the rear. The smooth discs would reduce drag by decreasing the skin friction over the wheels and by reducing the cross section as the front of the bike pushed through the air. It had a lightweight frame with oval steel tubing and bullhorn handlebars, putting his body in a position that would allow the air to flow efficiently over his back. He clipped his shoes to the pedals to increase power and wore a skintight suit covering his arms and legs. With his large support crew, he was well prepared for the altitude of Mexico City.

He posted a massive 50,808 metres, 1,377 metres further than Merckx, who was unimpressed: "It is the first time a weaker man has beaten a stronger man," he said.

Unbowed, Moser exploited the world's attention by repeating the feat four days later, this time with live TV coverage, and taking the hour record to 51,151 metres.

Despite its curved seat tube, the extreme downhill slope of the top tube and the steep seat stay, Moser's bike still had the same diamond configuration as other road bikes. The UCI allowed Moser's records to stand. Things went quiet for six years until the UCI dramatically changed the rules to allow new monocoque frames. These were frames that rejected the traditional safety bike diamond shape and used a single piece, usually made from carbon fibre.

There's an irony that the bike manufacturers of the early 20th century pioneered the production line and then dumped cycling to make cars. The Rover became the Rover car and James Starley's partner William Hillman created the eponymous Hillman brand (anyone remember the Hillman Imp?). This nearly killed off the nascent bike industry. It's nice, however, to find that one of the most iconic sports car makers, Lotus, should reverse things in the 1990s and take what it had learnt from making cars and apply it to cycling.

Lotus worked with Mike Burrows, the maverick bike designer. Burrows was the epitome of a cycling engineer: not only did he design bikes, he made and raced them. Some of his designs were recumbent bikes with only a single front fork: rather than slot the wheel between two forks, the wheel could be pushed on from one side. He realised that it would also have a drag saving: why have two drag-creating forks when you needed only one?

With Lotus, he used carbon fibre to create a single star-shaped piece to act as the frame with a single cantilever fork for the front wheel. All cross sections were narrow and aerodynamically profiled. The bike was demonstrated to Chris Boardman and his coach Peter Keen in a wind tunnel late in 1991 who immediately saw the aerodynamic advantages; data published later showed that its drag was about seven per cent less than a standard track bike. They worked to optimise the fit for Boardman who then used it to win the 4,000 metres individual pursuit at the 1992 Barcelona Olympics. It was Britain's first gold cycling medal for 72 years and the British press went nuts. The bike got as much press as Boardman did and was christened 'Superbike'.

By 1993, Boardman had announced that he would attempt the hour record. With only a week to go, news came of a relatively unknown rider called Graeme Obree attempting to do the same. Boardman must have been furious. Obree's bike was almost the antithesis of Boardman's high-tech machine: homemade, strange-looking and with a bearing and bottom bracket fashioned from an old washing machine. That's not to say it wasn't innovative. The front handlebars were T-shaped and set deep underneath Obree's torso so that his hands were tucked in under his chest. This odd-looking position was difficult to ride in but reduced drag by about four per cent compared to the bike Boardman would use. He'd also given it a name: 'Old Faithful', like a clapped-out horse.

Obree failed at his first attempt on the Friday night but, astonishingly, had another go on the Saturday morning. He beat Moser's record by almost half a kilometre and the hour now sat at 51,596 metres.

All Boardman could do was to continue with his own record attempt the following week as he'd planned. Using a bike every bit as innovative as Obree's, with a Corima carbon fibre frame and aerodynamic tri-bars, Boardman added over half a kilometre to Obree's distance; the record was now 52,270 metres.

The UCI didn't like Obree. Their opinion was that bikes should look like bikes – they didn't like Burrows' single-forked recumbent bikes and they certainly didn't like Obree's monstrosity. Quickly, they began to draft rules to ban Obree's bike: he responded by breaking the hour record again.

The UCI outlawed Obree's T-shaped handlebars but he easily got around them by opting for tri-bars similar to those already used in UCI races. As usual, Obree took his design to the extreme and made the tri-bars so long that his arms were stretched out in front. Ironically, the UCI's rule change had made his position even more aerodynamic. It didn't take long for this to be dubbed the 'Superman' position.

There was a fever to the hour record. Spaniard Miguel Induráin and Swiss Tony Rominger added another two and a half kilometres to it over the next six months before Boardman made another attempt in Manchester on 6 September 1996 using Obree's Superman position. On a modified Lotus bike with aerodynamic wheels, wearing an aero helmet and skintight Lycra, Boardman reached a world-shattering 56,375 metres. This record has never been beaten.

In 1997, the UCI changed the rules to the hour record again: they banned aero helmets, disc wheels, tri-spoked wheels, monocoque frames and anything that wasn't in their view 'traditional'. They wanted to turn the clock back to the purity of the Merckx era so that any hour record had to use the 1972 Merckx bike (drilling out the handlebars and chain and adding a titanium stem was conveniently forgotten). The new record would be called 'The UCI Hour Record'. Riding on anything else that wasn't Merckx-like, past, present or future was thrown in a bin and called 'Best Human Effort'. This is probably the greatest misnomer ever: the 'Best Human Effort' determines the best bike/rider combination while the 'UCI Hour Record' is the one that actually finds the best rider (presumably a human).

The model bike

I'll cut the UCI a bit of slack here for a moment: setting technical rules for sport isn't easy; I know this from sitting on the Technical Commission of the International Tennis Federation. The key is to understand the science of your sport at least as well as those you're trying to govern. If you don't, then every time something new is invented, you're scrabbling around to make a decision based on guesswork. The problem with the UCI was that they didn't seem to understand the basics of bike design which is why Burrows and Obree could sidestep them so easily.

Good governing bodies have ways of predicting how technology will affect their sport: tennis has 'Tennis GUT' and golf has its 'absolute distance rule'. I don't know if the UCI have such a model for cycling – I suspect they don't – but many other do. One of my PhD students – Richard Lukes – was a passionate cyclist and wanted to understand how technology affected performance. He created a computer model of the 250-metre velodrome in Manchester so that he could predict times in the 4,000 metres pursuit. As with the models made for bobsleigh and skeleton, it would give him an understanding of how to improve performance.

His model had the following inputs: the height and weight of the rider; the bike's weight; the rolling resistance of the tyres; the bike and rider's drag coefficient; and the efficiency of the frame, bearings and drive train. He could even change the altitude and the weather conditions.

The model worked by adding up all the forces slowing the rider down and subtracting them from the rider's force pushing the bike along through the cranks. It even took into account subtleties such as the changes in acceleration as the bike went around the curved parts of the velodrome track. It hadn't occurred to me until Richard's model that the rider doesn't go quite as far as the bike when going around a velodrome. This might sound nonsensical but the cyclist leans inwards as he or she navigates the curve at each end. This means that the top of the head traces a tighter circle around the track than the bottom of the wheels. The extreme would be a wall of death where a motorbike rider rides around the inside of a large wooden cylinder, going fast enough to be absolutely horizontal. The bottom of the bike travels faster and further than the top of the rider's head. For a velodrome, the difference is about three per cent.

The model can only work if the rider's propulsive force is known. This was something that could only be measured in a laboratory until special cranks were invented in the mid-1980s. The first 'power crank' able to work out on the road was pioneered by Ulrich Schoberer – he subsequently created a company called Schoberer Rad Messtechnik (Schoberer Wheel Metrology) or SRM. His cranks are wonderfully elegant and are one of the many things I wish I'd invented. Their current design uses four small strain gauges a few millimetres across attached to the main chain ring; this is the main cog at the front that the pedals are attached to. The strain gauges are calibrated in the factory to convert the small deflections when the cyclist presses down on the pedals into force.[98]

But cyclists aren't actually that interested in force, they're much more interested in power as this gives them an idea of how effective their bodies are at producing useful energy. Multiplying the force by speed gives power and this is shown either on a bike computer attached to the handlebars or downloaded later.[99] A typical value for an elite endurance rider is between 400 and 500 watts, enough to power my laptop, room lights and record player. From a standing start, they can briefly put out about 1,200 watts – enough to power a small radiator at the same time.

I gathered all the information I needed and put it into Richard's mathematical model, using Chris Boardman as my cyclist: 'virtual Chris' was 1.75 metres tall, weighed 68 kilograms and pumped out 442 watts at hour-race speed. The first thing I looked at was the tyres. A track bike's tyres try to do a similar job to the runners on a bobsleigh: they need to steer you in the right direction but with the minimum of friction. As the tyre rolls over the track, it flattens when it comes into contact with the surface and reshapes as it leaves. This takes energy out of the bike/rider system and is what causes rolling resistance. The contact patch with the track needs to be as small as possible, with just enough friction to stop the wheel slipping. The energy lost as the wheels roll over the ground can be minimised by using the largest diameter wheel possible, with thin tyres pumped up to the highest pressure.

I once explained this to an audience as part of a panel discussion when a Tour de France stage finished in Sheffield. It was held in front of 1,000 or so people at the Crucible Theatre (best known in sport for holding the World Snooker Championships). The audience weren't that interested in me or the other academics on the panel; we were just making up the numbers. They'd actually come to see Nicole Cooke, winner of the Olympic road race in Beijing in 2008, and David Walsh, the journalist who dared to challenge Lance Armstrong about drugs.

Although I'm a cyclist, I'm not a *real* cyclist. I've ridden around the Peak District National Park and around Majorca a couple of times, but that's about it. My role that day was to brief the audience about the science of cycling and I told them about the first bike designs and why bike aerodynamics is so important. I told them about the tyres.

"And don't forget," I said, pleased with myself, "pump up your tyres."

As I sat down next to Nicole Cooke I realised I'd probably just patronised her and everyone else with the art of the obvious. She was the professional, I was just an amateur; who was I to tell her to pump her tyres up? I sat there sick with embarrassment. Worse still, David Walsh announced that his pal Greg LeMond was in the audience. Three-time Tour de France winner LeMond waved shyly to the crowd which rose in unison to give him a minute-long standing ovation. I still blush.

So, my apologies to Nicole, to Greg and all the other great cyclists who might have been in the audience that day for teaching you to suck eggs. Here's a chance to redeem myself by taking Richard's model and perhaps tell you things that aren't so obvious.

The more the tyre deforms, even at a microscopic level, then the worse the rolling resistance. This means that thin tubular tyres made of silk, Kevlar or nylon are the best, knobby tyres are the worst and a smooth wooden floor is much better than Tarmac or concrete. Using Richard's model, I compared top-of-the-range Clement Colle Main tyres to the sort of cheap touring tyres you might get from an everyday sports store. The model told me that the cheaper tyres would have wasted about 20 watts of power and lost Boardman over 800 metres in his hour race. That's a one and a half per cent difference and well worth the extra cost of the fancy tyres.

When Eddy Merckx attempted the record, he focused on the bike's weight by drilling holes all over the place. It must be pretty important, then? Actually, not that much. The bike's weight only makes a difference when accelerating. This is important if you're going up and down hills, and during a road race you'll see the lighter riders zooming up a hill leaving the heavyweights behind, only for the heavyweights to catch them up again down the other side. But, since there are no hills in a velodrome (unless you go high up the bank during a lapse of concentration), then once you're up to cruising speed there are only minute accelerations during the bends. According to the model, adding a kilogram to the bike would change the distance ridden in an hour by only tens of metres.

I'm a little nervous about how insensitive the model is to bike mass, but I'm not the only one uncertain about the importance of mass as the reported bike weights of the hour record holders vary so widely: the lightest was Eddy Merckx's bike at 5.5 kilograms; the heaviest Fausto Coppi's at 9.5 kilograms.[100]

As with anything pushing through the air – bikes, golf balls, bobsleighs – aerodynamics is where the biggest gains are. David Bassett, from the University of Tennessee, gathered the best biking brains in the world to analyse the hour record.[101] All the researchers had at some time worked with their country's Olympic teams. (If you want the 300-page version, you should read co-author

Edmund Burke's seminal book called *High Tech Cycling*.[102] Sadly, Burke died of a heart attack doing what he loved – cycling.)

Bassett and his co-authors asked the question everyone wanted to ask: "If you put all the hour record holders on the same bike, who would win?" Would it be the technically-minded Moser or the maverick Obree? Would it be the powerful Induráin or would it remain the ultimate record holder, Chris Boardman?

In some ways, doing the maths for the model is the easy bit. The tricky part is getting information about the cyclists. Just what was the average power output of Eddy Merckx in 1972? What was Chris Boardman's height and weight in 2000? Different reports have his height consistently at 1.75 metres but his weight varies by at least two kilograms. While bike mass is not so important in the model, the rider's mass is because it affects the cross-sectional area – the bigger the athlete, the bigger the area and the larger the drag. The largest rider in recent years was Miguel Induráin at 1.88 metres and somewhere between 78 and 81 kilograms; the smallest was Tony Rominger at 1.75 metres and 62 to 65 kilograms. The essence of the model is that the cyclist with the best ratio of power to cross-sectional area is the one who'll win.

I'm going to assume that Bassett and his co-authors have the inside information on the athletes since they worked with a lot of them. The cyclists they chose were Merckx, Moser, Obree, Induráin, Rominger and Boardman, and they used a model similar to Richard's to put everyone on Chris Boardman's 1996 bike when he set the record of 56,375 metres at the Manchester Velodrome. Not only could they find out who was the best rider, but also how far they would go.

Boardman was the benchmark in their experiment with his distance of 56.4 kilometres (their model was only accurate enough to give distances in kilometres rather than metres). But how would Eddy Merckx do on Boardman's bike? His power was estimated to be lower than Boardman's so he finished behind him at 54.0 kilometres.

Out in front by a kilometre was Tony Rominger with a distance of 57.4 kilometres. His power was four per cent higher than Boardman's and he was three per cent smaller. He'd been disadvantaged in his own world record attempt with an inferior bike – my estimate is that it had 12 per cent more drag than Boardman's. On Boardman's bike, he travelled a full two kilometres further to take the win.

This kind of prediction is fraught with danger and I bet the authors had many arguments about their assumptions and results. Induráin's team only published the data of his 1994 record in 2000 and measured his average power output as 510 watts; Bassett estimated it as only 436 watts. Had they used this data, he would have come second rather than fourth.

When you have a model like this at your fingertips, it is always tempting to put yourself into it to see how you'd match up with the greats. I've cycled up the Sa Calobra climb in Majorca a couple of times: this rises almost 700 metres in ten kilometres with an average gradient of seven per cent. I know it's not the Alps, but the 26 hairpin bends are simply magical. My ride took about an hour and my tracking software told me I could sustain just over 250 watts. This seemed a bit high, but it's on the internet for everyone to see, so I'm going to take it.[103] I put my power output in the model and it predicted my own personal hour record as 38.7 kilometres. It's not huge but at least I'd have beaten Henri Desgrange back in 1892.

The UCI eventually saw sense and reversed their 1997 ban on technology in 2014. But the mark to beat wasn't Boardman's best record of 56,375 metres, it was the lesser 49,700 metres record set by Ondřej Sosenka on a Merckx-style bike in Moscow in 2005. Presumably they thought that Boardman's ride was unassailable. But the new set of rules have reinvigorated the hour record which has now been bettered five times: the current record holder is Bradley Wiggins with a distance of 54,526 metres.

The way the hour record grew shows how much performance can be influenced by innovative design. The coaches focused on reducing energy losses due to drag, friction, the bending of the bike components and on quantifying the power the cyclist could put into the bike using power cranks. My research team is currently working on ways to fit riders to bikes so that the cyclist is in the position that allows the highest power input to the bike; there's no point having a small bike that reduces drag if the effective power is also reduced.

The next chapter shows how biomechanists worked with designers in another sport to do the same by using technology to optimise power output. This transformed their sport within a few short years.

TEN

The flashing blade

CHRISTINE GLIDED slowly to a halt. She shuffled over to the side of the Calgary skating oval and gave me a wide grin: "It's really hard," she said. "I'm hardly getting any glide at all."

She was dressed like an old-fashioned Dutch schoolgirl, out for a leisurely skate on a frozen canal. She wore a bright blue woollen hat pulled under her chin, a dark woollen jersey, heavy cotton pants and a pair of old leather speed skates like something out of the sixties, which is exactly what they were.

Figure 35. Christine Nesbitt in 1960s (left) and modern skating apparel (right) at the Calgary skating oval. © *Steve Haake.*

Christine Nesbitt was a recently retired Canadian speed skater with a clutch of medals to her name. She had an Olympic gold in the 1,000 metres from the Vancouver Winter Olympics, an Olympic silver in the team pursuit in Turin and 16 other medals, half of which were gold. She also had a silver and a bronze in the World Allround Speed Skating Championships, a combined event where skaters compete in four different distances from 500 to 5,000 metres. You have to be seriously good to get a medal in a competition that ranges from sprint to long distance.

Christine had agreed with Kensington TV to try out some old speed skating gear to see how it performed. As a benchmark, she would compare herself to a 1960s skating legend called Inga Artamonova. She'd been a huge Soviet star in her day, starting out as a rower and then converting to speed skating. She was so good that she won the World Allround Speed Skating Championships four times between 1957 and 1965. Shockingly, she was murdered by her jealous alcoholic skater-husband in 1966.

Artamonova's best 1,000 metres was in 1963 with a time of 1 minute 35.00 seconds, giving an average speed of 38 kilometres per hour. In contrast, the current world record is 1 minute 12.18 seconds, almost 23 seconds faster. The question, as ever, was how much of this time was down to the equipment?

Get your skates on

Skating is older than you might think. While the ancient Greeks were sprinting naked in the hot midday sun down their lime-covered running tracks, hunters in the frozen north were wrapping up in furs and skating across frozen lakes to catch prey. Their skates were made from shaped animal bones strapped to their feet; these were probably used more for sliding rather than gliding and poles were needed to push the skater along.[104] This was skating for survival.

Wood replaced bone and by the 17th century skating had become a popular leisure pursuit, particularly in the Netherlands and Britain. One thing that helped skating grow was the 'Little Ice Age', a sustained period of cold temperatures that started around 1400 and lasted until the mid-1800s. The Thames regularly froze and much of the population of the Netherlands and eastern Britain seemed to need skates to get around. The first skating club – The Speed Skating Club of Edinburgh – was created in the 1740s while the first reported speed skating competition was held in the low-lying Fens of East Anglia in 1763. The golden age of Fen-skating was dominated by fantastically named skaters such as William 'Turkey' Smart and his brother-in-law rival, William 'Gutta-Percha' See.

The earliest wooden platform skates were strapped on to normal boots and poles were still needed for propulsion. In the Netherlands, however, the addition of an iron strip to the wooden skates changed everything. Friction between the blade and the ice was reduced during the glide and the sharp blade could be used to push off by angling the foot slightly outwards. The poles were discarded and the arms were now free to provide balance and poise. The thin edge of the blade was ground along its length to give it a hollow cross section like a W; this gave the blade two edges and allowed skaters to lean onto them when turning left or right. This technique became known as the 'Dutch Roll' and allowed skating speeds to increase.

The first all-steel skates were made in Philadelphia in 1850. Iron skates had been an advance but were heavy and lost their sharpness quickly; in contrast, steel skates kept their edges and were much lighter. Designs proliferated with beautiful curved fronts, cut-outs and embossing. It was around this time that the design of skates for dancing and speed skating diverged. Figure skates acquired a jagged toe pick that allowed the skater to suddenly dig into the ice and leap into the air; these exciting manoeuvres had exotic names such as Lutz and Salchow after the skaters who invented them.

The last thing that speed skating wanted, however, was a jagged toe with extra friction and the speed skate blade remained plain, simple and straight. The skate blade narrowed and the W-shaped cross section was dispensed with so that, close up, the edge of the blade was shaped more like a V. Rather than platforms to strap boots on to, the metal blades were attached directly to the soles using screws and straps. Only later did specialist ice skating boots appear with integrated blades attached by copper rivets. One way to reduce the weight of both figure and speed skating blades was to use a hollow tube as a beam along the top edge. Not only did this strengthen the blades but it also allowed them to become longer and thinner without bending during the high loads they experienced during skating. This helped Norwegian Axel Paulsen to create a new jump where he jumped off from his toe pick, spun in the air and landed backwards. It was named the Axel. The top tubes have now been lost on figure skates as stiffer steels have been used, but still remain on the much thinner speed skate blades.

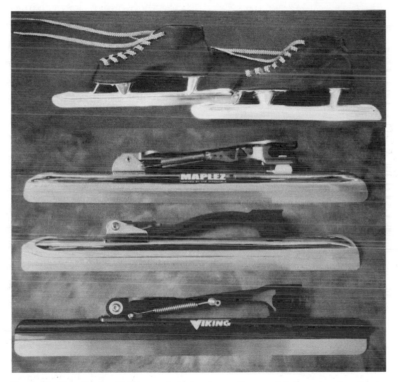

Figure 36. Top: Replica 1960s speed skating boots. Bottom: Modern clap skate blades with their hinged spring mechanism. The blades attach to the bottom of carbon fibre boots. © *Steve Haake*.

The first skating competitions took place on frozen lakes and canals, probably the most famous of which is in the Netherlands and is called the *Elfstedentocht*. This is a 200-kilometre race past the 11 historical towns of northern Friesland, starting and finishing in Leeuwarden. One of the key problems about skating on natural ice is the inconsistency of the weather, and the *Elfstedentocht* is only held when the ice along the whole route is at least 15 centimetres thick. In practice, this means weeks of sub-zero temperatures coupled with feverish expectation across the country as it waits for the official committee to announce the race. The *Elfstedentocht* is becoming a rare event: rising temperatures have meant that what used to happen every five years or so has happened only four times in the last 50 years.

One solution to the intermittent nature of natural ice is to create your own. Artificial ice ovals were first developed in London in the 1870s, although a *Glaciarium* had briefly existed as far back as 1841. It had been made of a rather smelly concoction of salt, copper sulphate and pigs' fat; it didn't last long.

Artificial ice is created by embedding pipes in concrete that carry a coolant pumped in from underneath so that it acts like a huge refrigerator. A thin layer of water is first put on top of the concrete onto which markings and adverts are painted, generally in blue and red, before successive layers are built up over it. Resurfacing the ice by hand was time-consuming and difficult but was revolutionised by the businessman Frank Zamboni. He patented a resurfacing machine in 1953 which was driven across the ice like a tractor; it shaved off the top layer of ice and sprayed on a thin layer of water. For the first time, Zamboni created a uniform skating surface that could be replicated wherever and whenever it was wanted.

The first artificial speed skating oval was built in Gothenburg in 1953, but it took until 1990 for all top-level competition to be on artificial ice.[105] Even then, the first *indoor* artificial oval wasn't opened in East Berlin until 1986. The effect of these new arenas was to help create better and more uniform ice and shelter both the spectators and the competitors from the elements. As I stood on the translucent ice in the Calgary Oval, I could see right through the polished ice to the painted lines a finger's depth below. It was so clear I expected a shoal of fish to flutter past.

The measure of a surface's roughness – or even the opposite, its slipperiness – is the coefficient of friction. As an example, if you weighed 80 kilograms, then your downward force on the ice would be about 800 newtons. If it took a force of eight newtons to get you moving, then the coefficient of friction would be the ratio of the two or 0.01. Flat surfaces generally range between zero and one[106] and the coefficient of friction of skates on well-prepared ice is actually only about 0.004; this means that a push you might give to make a doorbell ring would get you moving.[107] A concrete surface next to the ice has a coefficient of friction of almost one so you would never slip on it unless it was wet, when the coefficient of friction would drop by half due to the lubricating effect of the water. Our eyes and brain are very good at judging what to expect when we are walking and it's only when we get an unexpectedly low coefficient of friction that we fall over. The slapstick banana skin springs to mind.

The size of the coefficient of friction depends upon the nature of the two surfaces coming into contact, so it's not just the ice but also the blade that is important. Because of this, skaters spend an inordinate amount of time sharpening their blades with special tools they take to tournaments. They even bend the blades slightly so that they follow the direction of the oval. As a non-skater, it's hard to imagine this could make much of a difference but skaters clearly do. In 2013, US short-track skater Simon Cho was given a two-year suspension for cheating after confessing that he'd intentionally damaged a rival's blades. He'd sneaked into the changing rooms and used a skate bender to distort Canadian Olivier Jean's blades during the 2011 World Team Championships. The gamble failed: Canada got a bronze while the US failed to get into the medals.

Having a thick skin

In the mid-1970s, clothes were generally bright, wide and voluminous, and flared trousers were the height of fashion. Skaters still wore traditional heavy sweaters, leggings and hats. So in 1974, when Franz Krienbühl stepped out onto a skating oval in a skintight suit, people laughed – it hinted at desperation. He was usually a back-marker during races, and, having been born in 1928, he seemed too old for such a revealing suit. But the laughs turned to uncertainty and then envy as Krienbühl's times in the 10,000 metres dropped by over 40 seconds. In the next two years before the 1976 Winter Olympics, all speed skaters switched to the suits and Krienbühl, perhaps with an advantage as an early adopter, managed his best 10,000 metres ever, coming in eighth. Having decided to take up speed skating at the age of 38, Krienbühl still had time to become Swiss champion 14 times, the last when he was 55.

Manufacturers keen to exploit the skinsuit craze tried to improve on Krienbühl's first Lycra suit. One of the first experiments with speed skaters in a wind tunnel was done in 1981 by Gerrit Jan van Ingen Schenau from the Free University of Amsterdam. He put six male speed skaters in a large wind tunnel capable of speeds up to almost 70 kilometres per hour.[108]

As we saw with golf balls, the drag is affected by the type of flow in the boundary layer at the surface, and the faster the air speed, the more turbulent the air in the boundary layer and the lower the drag. The dimples on a golf ball brought this transition down to speeds that could be achieved by a golfer. Van Ingen Schenau's experiments revealed that the drag crisis for speed skaters was already in the range at which skaters skate, between four and 12 kilometres per hour. But, since a skater's legs and arms moved through the air faster than the torso, they would reach drag crisis speeds while the torso might not. This was a manufacturer's dream because panels of different fabrics and roughnesses could be used on different parts of the body to optimise drag reduction. And since a 500-metre speed skater travelled faster than a 5,000-metre skater, the optimum suit for each distance would be different. This variety of options meant that manufacturers could evolve suit design for different distances, declaring the latest suit even better than the last.

Nike, Descente, Hunter and Mizuno brought out new suits prior to the 2002 Salt Lake City Winter Olympics. Rough textured fabrics tended to be used on the arms and legs with smoother polyurethane-coated fabrics on the torso and thighs. Hoods were introduced which became contoured to stop bad airflow around the neck. Low friction fabrics were put on the inner thighs, although this had nothing to do with aerodynamics. Most speed skaters have really impressive legs due to all the training they do and the inner thighs tend to rub together. A nice slippery fabric reduced the coefficient of friction and allowed them to slide easily past each other.

Lars Saetran and colleagues from the Norwegian University of Science and Technology tested six different suits on a mannequin in a wind tunnel. It was clear that the manufacturers had different ideas of what was optimum. One was super-smooth on the lower legs while another was super-rough; the four others were somewhere in between. Two suits were completely smooth along the arms while the other four had texture. There were only two things that were consistent. Firstly, the torso was always smooth. Secondly, each manufacturer claimed *they* had the best suit.[109]

Of course, you won't necessarily win, even with the best suit. Christine was interviewed after the Canadian team switched to the new Apogee suit before the 2014 Winter Olympics in Sochi and said: "The suit doesn't skate well for you." The Americans should have heeded her words. Their suit at the Games was made by Under Armour Inc. in collaboration with Lockheed Martin, the aerospace manufacturer. They worked together in secret to create a suit that would beat the best and, two weeks before the Games, launched the Mach 39 skinsuit, calling it 'the fastest speed skating skinsuit in the world'. It was introduced with an exuberant American fanfare.

What Under Armour failed to do, however, was work properly with the athletes themselves. They kept the new suits under lock and key and the first time most athletes used them was just before the Games. The Mach 39 story could have gone two ways. If the Americans had won events from the outset, the suit would have been hailed as a blinding success, the athletes would have revelled in their technological superiority and they might have gone on to sweep the board. Alternatively, if the Americans started badly, then the suits would have been called into question and doubts might have crept into the minds of the athletes, with thoughts they'd been given a dud by nerdy engineers who knew nothing about speed skating.

Which way did it go? The American speed skaters failed miserably at Sochi, failing to win a single medal for the first time since 1984. The Mach 39 was panned by the athletes, the press and the public.

Hanging loose

Christine showed me her modern gear. Her shiny suit was a tight-fitting one-piece affair from Apogee with a contoured hood as tight as a swimming cap. As befitting an elite athlete, she wore a cool pair of shades in sunflower yellow. She agreed to do a full-out 1,000 metres, going as hard as she could. Her time of 1 minute 21.00 seconds gave her an average speed of 44 kilometres per hour. Although this was over eight seconds down on her 2012 world record when she was at her peak, this was still very respectable for an athlete who'd been retired for three years.[110] Where I live, travelling that fast would get you a speeding ticket.

As she shot past, her skates gently clacked as if a component was loose. Rather than have blades fixed to the bottom of the boots, they pivoted about the front foot so that they hung down at the back. On each stride, the blade dropped down only to be pulled back upwards by a spring, making the clacking sound I'd heard. Why would you make a skate so loose that the blade hung down? Surely skating is hard enough without a dangerously loose component?

I've skated around my local rink and I'd never been told how to skate fast properly: I'd just assumed it was a bit like running. When you run, you swing one leg forwards in the direction of travel and push off against the ground with the other foot. The coefficient of friction between a running shoe and a surface is really high at around 1.0, and the foot doesn't normally slip. Running on ice is almost impossible because as you push off, your foot just slips backwards.

Ice is both good and bad. On the one hand it's good for gliding, but on the other hand it's so slippy that it's difficult to push off. I watched Christine as she effortlessly glided past with her long, flowing strides. Her trick was that she didn't push backwards but outwards at right angles. As she pushed out, the blade dug into the ice slightly; as she moved forward, this foot was soon retarded behind her and, while the push had been initially outwards, she was now pushing backwards, propelling herself forwards.

Elite skaters spend years perfecting their technique and developing the right muscles for it. I saw a line of speed skaters training on the steep steps to the upper seats of the Calgary Oval. Each step was at least 30 centimetres high. They would jump onto the first step, with their backs to the rink. They would crouch with their upper body almost horizontal and with their knees bent, straining until their coach shouted them to jump again, not *down* the steps but *up* them, to hold the crouch again. I was in agony just watching.

The problem with pushing off at right angles to the direction of travel is that it's inefficient in terms of human body design: it's just not what our leg muscles were designed to do. Our natural approach to locomotion is to stretch a leg out in front so that it's still a little bent when the foot hits the floor. The body moves over the top and the leg swings backwards; then, in the push-off, the leg

183

is extended and the ankle flexed with the toes the last thing to leave the surface. A good push will bring your heels upwards towards your backside.

This method doesn't work in speed skating. If you try to extend your ankles so that you push off with your toes, all that happens is the long blades dig into the ice: the best case is it slows you down, the worst case is you fall over. The upshot is that with a fixed blade, the skater has to keep the ankles stiff to stop them being used in the push-off. You can gauge how much of a hindrance this is if you imagine jumping up and down on a trampoline. If you let yourself naturally extend your toes as you push off the elastic bed, you jump really high. If you keep your ankles rigid so that your feet are at right angles, your bounces are boringly low. This is the effect the old fixed ice skates had.

This lack of ankle extension was a problem since speed skaters had to learn to check this action through years of training, extending their legs and ankles just enough, but not by too much. Gerrit Jan van Ingen Schenau was looking at the mechanics of speed skating at the Free University of Amsterdam and found that the unnatural technique was making skaters' shins painful. He suggested something radical: release the blade at the heel so that it remained in contact with the ice while the boot tilted forwards about the toe. This would allow the skater to extend their toes and relieve the pain. It might also improve performance.[111]

The first versions of these new skates were made in 1984 and used in a 500-metre race. They were nicknamed the klapskate – not because of the noise they made but because *klappen* means 'to slap on' as in 'to slap on a little bit extra'. It was soon anglicised to clap skate. Other skaters watched on with interest, but were reluctant to use them for fear of ruining their hard-earned technique.

The Dutch researchers joined up with a manufacturer called Viking to apply for a patent. Surprisingly it was rejected because the clap skate had already been invented almost 100 years before by Karl Hannes of Burghausen in Germany. His intention had been the same, to make skating easier by putting a hinge at the toe and allowing the blade to dangle at the heel. Evidently it hadn't been well received back then either.

Despite the patent setback, the researchers continued with their innovative design, trialling it with a few skaters who agreed to learn how to stretch out their ankles during skating and try to forget everything they'd been taught previously. This was the only way they could take advantage of the clap skate design. The problem was that in the pressure of competition, the skaters reverted to their old technique so that they might as well have been wearing their old skates.

Two more PhD students joined the speed skating research team at the Free University. Jos de Koning and Erik van Kordelaar were both speed skaters and were convinced by the new clap skate design. Gerrit and Jos would present to speed skating coaches and teams on how the drop-heel design should improve performance. While they were greeted warmly and even agreed with, skaters still didn't change. Erik took the lead in the 1990 season and swapped over to the skates. His times improved so much that by the 1992 season he was skating three per cent faster. His new self on the clap skate would have beaten his old self on the fixed skates by nearly 17 metres over a 500-metre race.

The breakthrough came when Erik became the coach of a junior skating team in the south of Holland and convinced the whole team to switch skates. This must have been both an exciting and anxious time for them all: it could have gone horribly wrong and their performances for the year would have been ruined. They took 11 junior skaters and gave them the new clap skates, tracking their performance and comparing them to a similar group of skaters on conventional skates to act as a control. The results were startling. The skaters on standard skates improved by two and a half per cent: the clap skaters improved by over six per cent.[112]

This was still not enough to persuade the elite squads to switch, particularly as they were now worried that the International Skating Union might ban the skates. By the winter of 1996, five years after Erik van Kordelaar had used the skates for himself, three elite athletes finally used the skates in competition. The Dutch women were in the vanguard and one by one they changed to the clap skates. Then, Tonny de Jong won a gold at the European Championships. The International Skating Union saw the future and were happy with it, taking the bold decision to approve the clap skates. The men finally tried them in training and, when they saw that world record times were possible, they switched to them too.

By the 1998 Winter Olympics in Nagano, every speed skater used the new clap skates and every world record was broken.[113]

The clap skate effect

I asked Christine if she would like to compare the old boots to her current state-of-the-art ones and whether she could beat Inga Artamonova's time for the 1,000 metres of 1 minute 35.00 in them. She thought briefly and grinned 'yes'.

Modern speed skates are beautiful things. They are made by taking a plaster cast of the athlete's feet which are then used to mould a pair of tight-fitting carbon fibre outsoles to fit the blades to. Old-fashioned boots are simpler, made from soft leather with the fixed blades screwed to the sole. She agreed to wear 1960s clothing rather than her skinsuit, sourced from a museum in the Netherlands.

"Go to the start," shouted the starter.

Christine took up an aggressive pose in her old-fashioned gear, one foot pointing forwards, the other planted firmly sideways behind her. Her friends yelled encouragement.

"Ready."

The gun went off and Christine launched herself forwards, pushing off from the angled blades with a splatter of ice. She accelerated into the first turn, elegantly crossing her feet over as she cornered. The split time for the first 200 metres came up on the board: 21.43 seconds. Her friends screamed support and I noticed clouds of ice shards bursting upwards on each stroke. At 600 metres her time came up on the board as 56.40 seconds and I could see that she was labouring. I tried to work out if she could beat Artamonova, but my brain didn't seem to function in the excitement: 30-something seconds to go.

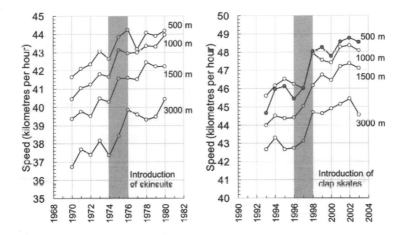

Figure 37. Best performances in long track speed skating showing the effect of the introduction of skinsuits and clap skates. *Data courtesy of Leon Foster.*

She raced down the back straight, the thick woollen top and the tag of her trousers flapping in her wake. She turned the last corner and I glanced at the clock: it was going to be close. She raced towards the line, pushing one leg forwards and everyone turned to the clock: 1 minute 34.88 seconds. Christine had beaten Artamonova's time by just a tenth of a second – had they been racing together, Christine would have won by just a metre.

Taking away Christine's modern gear increased her 1,000-metre time of 1 minute 21.00 seconds by 13.88 seconds. Conversely, the introduction of the skinsuit and the speed skates must have improved her time by the same amount, or 15.7 per cent. This equates to an increase in speed of about 6.5 kilometres per hour.

As one of the first athletes *never* to have raced on fixed skates, Christine clearly struggled to stop her foot tilting forwards which is why the blade dug into the ice and shards sprayed everywhere. Since each stroke was less efficient she speeded up her stride rate; a quick count off the video showed that it increased by about eight per cent.

Research shows that the 500-metre sprinters didn't get quite as much out of the skates as those doing the 3,000 or 5,000 metres, probably because more time was spent accelerating at the start compared to the time in cruising mode.[114] Women benefited more than men which is perhaps why they took it up more quickly. A look at historical data, however, shows that the benefits from the skinsuit and the clap skate were up to three kilometres an hour each. This makes them some of the biggest technological advances in sporting history.

I love the idea of the clap skate because it was so counterintuitive. Like Baron von Drais' two-wheeled velocipede, the clap skate came from a light-bulb moment combined with sheer dogged belief that the idea was right. History shows they were. The next chapter features similar pioneers who created a whole new genre of sport, and then invented new technologies to go with them.

ELEVEN

Superheroes

15 JULY 2007. I was sitting in the wet grandstand of the Don Valley Stadium in Sheffield. The rain was lashing down and, recently, the city's rivers had overflowed after a month's rain had fallen in a single day. Two metres of floodwater had divided the city and three people had died. Tonight was light relief from the city's problems and we'd come to see the Norwich Union British Grand Prix Athletics. Maybe the cheers were more effusive than normal to make up for the awful conditions – we were just thankful that the athletes had still agreed to come. There was one person we were anxious to see: Oscar Pistorius.

Oscar was the poster boy of para-athletics and had the coolest of nicknames: 'Blade Runner'. As a double below-knee amputee, he ran on two carbon fibre legs shaped somewhere between a 'C' and an 'L'. The stumps of his legs slid into carbon fibre sockets with the blades attached below. The reason tonight was special was because Pistorius had recently received approval from the IAAF to run in able-bodied races. So here he was, a double amputee competing against able-bodied runners. But it wasn't approval as such; confusingly, it was merely not disapproval. The IAAF had recently put in place rule 144.2 to prohibit 'technical devices designed to improve performance'. Most people had read this rule as a euphemism for 'that means you Oscar' but after dogged questioning by the media, the IAAF had said he could run until they could work out whether his blades gave him an advantage or not. So, a Schrödinger's Cat of a decision: not approved and not banned at the same time.

The Sheffield race was actually his second 'able-bodied' race, as he'd run one a couple of days previously in the magnificent surroundings of the 1960 Olympic Stadium in Rome. He'd run his usual 400-metre race, coming from behind to take second place with a time of 46.90 seconds. And now here he was in soggy Sheffield, surrounded by the flotsam and jetsam of the flood, staring at the rain bouncing off the track. Our expectations were unreasonably high.

The race was the last of the night, just before nine o'clock. The runners were introduced and Pistorius got one of the biggest cheers. They went to their marks, got set and bang went the gun. One of the runners immediately stumbled and came to a stop, hands on hips. Was it Pistorius? No, he was out in lane eight, the lane he hated most. It was someone in the middle, Jeremy Wariner in lane four, the fastest 400-metre runner in the world at the time. At the 100-metre mark, Pistorius was behind and the crowd cheered wildly with the expectation of seeing him come through as he normally did. But something was wrong and he looked like he was running through treacle as he rounded the final bend – it was probably water. He looked a dispirited figure as he came in last, 20 metres behind the leader with a time of 47.65 seconds. To make things worse, he was disqualified for running outside his lane.

The press weren't particularly interested in his performance. They were far more interested in his desire to run in able-bodied races. Why did he want to run at the Olympics? Weren't the Paralympics good enough for him? More importantly, were those carbon fibre blades an unfair advantage? Were they cheating?

Figure 38. Oscar Pistorius in the 2011 World Championships in Daegu. © *Erik van Leeuwen.*

In the beginning

The fact that this discussion was going on at all would have been remarkable for Paralympic athletes just a few decades before. As late as the Second World War, those with spinal injuries returning from battle were considered lost causes. You were lucky if you survived for more than a few years. Athletics? A mere dream.

But wherever there is war, there is innovation and there have always been examples where someone with vision has refused to see disability as life-ending. There are stories of soldiers from the third century BC with leg prostheses made with a wooden core and a copper or bronze outer sheath. We've probably all seen the stereotype of a pirate with a wooden leg and hook for a hand: that was state-of-the-art for the Middle Ages. In the 15th century, a mercenary in Germany had his hand shot away. He had it replaced with an iron one, giving him the name *Götz von Berlichingen mit der eisernen Hand*. Although the moniker wasn't as punchy as Blade Runner, both it and his iron hand allowed him to continue his reign of terror for another 40 years.

Things became more sophisticated in the 16th century when Ambroise Paré from France travelled from battle to battle tending wounded soldiers. He created an artificial leg with an articulated knee and a prosthetic foot not too dissimilar to those today; he also made a mechanical hand operated by springs. The occasional wheelchair made an appearance with one creation in 1655 by watchmaker Stephan Farffler from Nuremberg having three wheels and a hand crank for propulsion.

In 1815, the Earl of Anglesey had his leg blown off at the battle of Waterloo. He might have resorted to the ignominy of being pushed around in a bath chair, the wheelchair of choice for invalids of the time, but instead he had a fully articulated leg made. It had an upper leg and socket made of wood and a steel knee joint with internal 'tendons' made of tennis strings attached to the foot. Flexing the knee made the foot flex so that the Earl could walk in a realistic manner. However, the clanking noise it made when he did gave it the nickname 'the clapper leg'. Variants of the clapper were still on sale during the First World War when thousands of soldiers returned maimed from the front.

Companies such as Blatchford in the UK began to innovate and, by the Second World War, had legs that would swing naturally and quietly during walking. This coincided with a small revolution that was taking place in a hospital outside Aylesbury. In 1943, the British Government asked a renowned doctor who'd fled Nazi Germany to take on the task of looking after soldiers with spinal injuries returning from the war. He agreed but with the proviso that he did it *his* way. They agreed and Stoke Mandeville received Ludwig Guttmann on to its staff.

'Poppa' Guttmann was firm in his opinions and his management style: to some he was a dictatorial headmaster, to others an inspirational father figure. His first innovation was to treat those with spinal injuries as people with a future rather than those who would die and so weren't worth bothering with. His next innovation was to introduce sporting activities into his curriculum, using it as a means for both physical and mental rehabilitation. Archery was his first choice of sport as it involved upper-body strength and balance and could be attempted by almost anyone in a wheelchair.

Probably the seminal moment for Paralympic sport came on 29 July 1948 when Guttmann held an archery competition between Stoke Mandeville and a disabled team from the Star and Carter Home in Richmond. Guttmann was no fool; he'd purposely selected the opening day of the 1948 London Olympics. From the start, his ambition had been for his games to become the equivalent of the Olympic Games for the disabled. It became a yearly event and, by 1956, 18 different countries came to the games.

Because of its origins in the spinal unit, sports in these early tournaments were primarily wheelchair-based. Archery was followed by wheelchair polo which was quickly found to be too violent and was replaced by wheelchair basketball. My first experience of this sport was in 1977. I was 13 and it was my first trip abroad on my school's French exchange to Brittany. I stayed in a seaside town called Lorient and I remember just two things about it. The first was an enormous concrete U-boat base left over from the Second World War nestling at the end of the bay. The base was impregnable to aerial bombardment so the city bore the brunt of the bombs instead. The second was a basketball game a short drive away up the coast. My French was rudimentary so I was surprised to find when we arrived that all the athletes were in wheelchairs.

The team must have been the Club Olympique de Kerpape, two-time French champions. I remember being fascinated by the energy and violence of the game; these athletes were smashing into each other and clearly loving every minute of it.

In the early days, players would still use the early *travaux* (works) wheelchairs which were more like huge padded armchairs with wheels attached. Terry Willett, who captained the British wheelchair basketball team around that time, described how players would intentionally use the dangerous edges of their wheelchairs to take out other players.

Figure 39. Early *travaux* wheelchairs used for hockey. *Image courtesy of the International Wheelchair and Amputee Sports Federation, National Paralympic Heritage Trust and National Spinal Injuries Centre at Stoke Mandeville (Buckinghamshire Healthcare NHS Trust). © NSIC.*

They were not particularly fast, but dangerous in the right hands. Willett tells how the Australian team used the *travaux*'s handling to intimidate players during the 1970 Commonwealth Games final in Edinburgh.[115] Because the castors were at the rear, they could turn on a sixpence and the Australian player Mather Brown would turn his chair 'accidentally' to catch the knuckles of his opponents. British playmaker Cyril Thomas was targeted repeatedly and, finally, as blood began to drip from his knuckles, he warned Brown that he'd get flattened if he did it again. Of course, he did it again and Thomas 'laid him one, right on the nose, knocked him out cold'. The Australian was carried off, the Brit was sent off, but the British got the gold.

Most athletes had a single chair that was used for everything. It seems odd to look at old photos of a standard chrome Ministry of Health chair being used in a basketball game; 25 kilograms or more in weight and complete with unnecessary push handles sticking out of the rear. The best players could do to get an advantage was to sit on cushions and put their feet on blocks to raise their height.

Chariots of fire

Basketball wasn't the only form of activity possible in a wheelchair and it was inevitable that people would try racing each other. Guttmann maintained that paraplegics should be limited to 60-metre races because of their low blood pressure and the likelihood they would collapse.[116] The Ministry of Health wheelchairs of the time didn't help the budding athletes: they were heavy, had thick tyres and tiny castors out front. Standard issue was the collapsible wheelchair made by Jennings. Constructed out of tubular steel, they were lighter than the old *travaux* chairs but still weighed a hefty 20 to 30 kilograms. With more or less a monopoly in the wheelchair market, Jennings didn't care to innovate until the late 1970s when some racers, frustrated with the one-size-fits-all approach to wheelchair design, began to build their own. There was

Bosse Lindquist of Sweden, Peter Carruthers and Paul Cartwright of the UK, Rainer Küschall of Switzerland and Bob Hall of the United States. All began to attack the lazy aspects of wheelchair design: poor fit, high rolling resistance, overly engineered and heavy.

Bob Hall was the first official wheelchair athlete in the Boston Marathon in 1975. In an un-politically correct move by modern standards, the race director had stipulated that Hall could only get an official time if he beat three hours, otherwise he was just someone in a wheelchair who'd wheeled around the course. Hall beat it by two minutes. An ex-Stoke Mandeville patient called Rainer Küschall began to design chairs in Switzerland. He'd been cajoled and inspired by Guttmann and, as a quadriplegic with poor arm function, aspired to be as manoeuvrable as paraplegics (those with injuries to their legs but not their upper body). He realised that the chairs he had were pretty useless and began to strip away things that weren't needed. He replaced the footrests with a textile strap and saved ten kilograms immediately. Soon he was using aluminium to reduce weight further. Importantly, Küschall realised that he had to start with the person and build the chair around them.

The wheelchair repair tent at the 1984 International Stoke Mandeville Games rapidly became a show tent for new designs, with innovations making their way back into everyday wheelchairs. The manufacturer of the clumsy Jennings chair went into bankruptcy.

Paralympic sport was still in its infancy but the men's 1,500 and women's 800-metre races became demonstrator events at the 1984 Olympic Games in Los Angeles. Watching these events now is like watching wheelchair evolution in action;[117,118] a journal paper around that time said trying to describe a racing wheelchair was like 'hitting a rapidly moving target'.[119] By 1984, women's wheelchairs no longer looked like they belonged in a hospital. They had two large driving wheels either side of the seat with rubber-covered push rims. All but one had the standard two castors low down under the front feet. Candace Cable's chair stood out: it had two large spoked wheels at the front about half the diameter of the main wheels and with thin tubular tyres. A few of the athletes wore gloves to help them push the rims.

The main difference between these racing chairs and their ancestors was that the main wheels were not vertical but had a camber. This meant that the bottom of the wheels in contact with the ground were further apart than the top. It had a number of benefits. Firstly, it made it easier for the athlete to push down on the rims and bring the arms back up for the next push. Secondly, the camber helped stability during cornering. When turning, an athlete experiences a sideways acceleration much as you do in the inside of a car on a bend. If this is too high, the wheelchair tips over. Angling the wheels inwards counteracts these accelerations in the bend: the larger the camber angle, the larger the counteracting forces and the faster the athlete can turn.

The Seoul Olympics four years later also had wheelchair demonstrator events and evolution was clearly in action. Three of the eight women in the 800 metres had wheelchairs with only a single half-diameter spoked wheel at the front. Others still had two front wheels, some big and some small; Ann Cody-Morris' two castor wheels were so close together that they might as well have been one. In 1984, they'd clearly noted that the three medallists had worn gloves: this time they all did.

As in cycling, racing wheelchairs need to have large wheels with well-pumped-up tyres to keep the rolling resistance low. Of course, another way is to have fewer wheels. The message in Seoul was clear: the medals all went to those in chairs with three wheels; the four-wheelers trailed behind by a massive ten seconds.

By the time of the Barcelona Olympics and Paralympics in 1992, all athletes used wheelchairs with three wheels. A previous rule on the maximum length of the chairs was relaxed and frame designers took advantage by moving the front wheel to the end of a long beam. A simple steering mechanism could be tapped once to give the exact turning circle of the bend and back again in the straights. Between Los Angeles in 1984 and Barcelona in 1992, wheelchair racing improved so much that times in the women's 800 metres dropped by 15 per cent. In 1984, the difference between the first and last athlete was 28 seconds; by 1992, it was down to just two. Both innovation and competition had become fierce.

Late in 2001, my team and I were asked by UK Athletics to investigate how racing wheelchairs could be improved for the next Olympics in Athens in 2004.[120] We timed the British athletes in sprint starts using every timing gate we could lay our hands on. We weighed them and their chairs and did rolling resistance tests on one of Tanni Grey-Thompson's chairs. We sat it on the belt of a huge treadmill and attached a cable to its front as if it was to be towed. Instead, we ran the cable over a frictionless pulley and hung a small set of weights from it.

When the treadmill was switched on, the wheelchair started to move backwards, but was held in place by the weights pulling down on the cable. The chair's wheels rolled over the belt passing by underneath. The measurement was simple: the rolling resistance of the wheels was the same as the force of the weights pulling down on the cable. Disaster struck during a later test when the cable broke and a chair flew off the back of the treadmill, tearing the belt and doing thousands of pounds' worth of damage. We weren't popular.

We played around with the camber to see how it would affect rolling resistance and found it was lowest when the wheels were eight degrees from vertical. This would still drain about 20 watts of power from the athlete at top speed, about a fifth of their total output. The rolling resistance was really sensitive to the angle of the wheels as it was easy to have them pointing 'toe-in' or 'toe-out' when setting them up; getting it wrong could easily double the rolling resistance.

Two of the engineers on the project, Terry Senior and Nick Hamilton, designed a portable alignment tool that could be used at the side of the track, copied from those used for racing cars. Two spirit-level lasers like those used by architects were attached to the wheels with a further one on the midline of the chassis and oriented to shine forwards onto a small screen. If the wheelchair was set up right, then the lasers would shine a vertical midline and two wheel lines looking like this: / | \ . Adjusted correctly, the lines for the wheels would be about eight degrees from vertical and equidistant from the midline. It was simple and elegant and won an award at a sports fair in Germany.

With our knowledge of cycling, we also investigated the chair's aerodynamics. The funding wouldn't pay for the use of a full-sized wind tunnel, so we used computational fluid dynamics instead. We split the chair and athlete up into their constituent parts to work out the contribution of each to the total drag. Fifty-eight per cent of the drag came from the athlete, 17 per cent came from the rear wheels and the remaining quarter came from the frame and seat. We had a team of about half a dozen working on the project from the two universities in Sheffield and were pretty pleased with our report: we made recommendations on body position, wheel alignment and athlete training. But, elite sport is pretty ruthless and there was always a tension in meetings with UK Athletics that I never understood. The project didn't continue.

I asked Nick Hamilton what he thought we did wrong. "Maybe sometimes you just answer the wrong question," he said. I never did find out what the right question was.

Figure 40. Key components of a wheelchair racer for a computational fluid dynamics model to calculate aerodynamic drag. © *John Hart.*

After this, other teams did their own testing and developments and big companies such as BMW and Honda got in on the act. In the same way that bikes developed, the frames were made from aluminium, titanium and then carbon fibre. The front wheel moved further forwards and was carefully balanced so that most of the weight was on the larger rear wheels, reducing the rolling resistance on the front. Watch a race and you'll see the chairs bobbing slightly off the ground as the athletes push; with the same principle as bobsleigh and cycling, the front wheel needs to touch the ground just enough to steer and no more.

Flex your feet

Prosthetic limbs experienced a similar revolution to wheelchairs between the Second World War and the Seoul Paralympics in 1988. Up to the 1980s, the main prosthetic for amputees was the Solid Ankle Cushioned Heel or SACH: this had a rubber heel with a wooden insert and a single pivot representing the ankle. Its purpose seemed to be to look like a foot rather than act like one and it was heavy and cumbersome. One of the problems with the heavy prosthetic was that it increased the moment of inertia of the leg in the same way that the *halteres* did in Chapter 2. The extra weight in the ancient long jump might have been good for increasing the force during the take-off, but if you wanted to run fast, the last thing you wanted was heavy weights for your feet. Another problem with the SACH was that, unlike a real foot, it absorbed a lot of energy during footfall and gave very little of it back.

In 1975, an active 20-something student called Van Phillips lost a leg in a boating accident. He used a SACH foot and hated the thing. He decided there had to be something better and by the early 1980s had started to tinker with carbon fibre footplates that were both lighter and more responsive. He would make his own carbon fibre feet and play tennis on them until they broke. In the winter of 1982 he met an engineer called Dale Abildskov from the University of Utah and they hit it off so well that, within three weeks, they'd designed and built a carbon fibre blade shaped like a 'C'. Phillips ran down the hallway of Abildskov's condominium and was elated; it was fast, it was bouncy, it was just like running again.[121]

The prosthetic acted as a leaf spring in a similar way to those on truck axles. To be effective as a prosthetic, however, it had to be able to store and return the maximum amount of energy for the lightest possible weight. This required a material that didn't break at high stress, was flexible and had a low density. Carbon fibre, as Phillips had figured out already, was the only real choice. He called his new prosthetic the 'Flex Foot' and set up a company with the same name.

The first Flex Foot on the market was shaped more like a J (if you were running right to left) and had a flat horizontal plate sticking out of the rear to mimic the heel. The J was loaded during running so that it compressed by several centimetres and returned to shape just as the runner pushed off from the toe; it gave back about 80 per cent of its energy to the runner.[122] Phillips' invention got off to a great start in its first Paralympics in Seoul in 1988, with Dennis Oehler winning three gold medals in the 100, 200 and 400 metre sprints.

About the same time, down in South Africa, a small blond-haired toddler was just beginning to walk. As with any child, his parents thought he was special: he was. Little Oscar Pistorius had been born with bones missing in his lower legs and, when he was 11 months old, his parents took the agonising choice to amputate, feeling that this would give him the best chance of being able to walk with prosthetics. He got his first set of legs when he was two, with the lower part made of heavy wood and rubber. Despite being, in his own words, 'a real menace', Oscar's parents encouraged him to ignore his disability and live life to the full.[123]

By 2003, Pistorius was a typical testosterone-filled 16-year-old high-school student, playing practical jokes on his non-disabled classmates and throwing himself into any sport he encountered. He used a prototype prosthesis designed by Chris Hatting, an engineer friend of his father. When Pistorius was seriously injured playing rugby later that year, he took up sprinting as part of his rehabilitation. At a local event in January 2004, he easily won the 100 metres in a time of 10.72 seconds; to everyone's surprise, including himself, it was an unofficial world record.

Like pieces of a jigsaw coming together, Hatting was headhunted by Van Phillips' Flex Foot which had been bought out by an innovative Icelandic company called Össur. Pistorius flew to California in June 2004 and was fitted with their very latest product – 'Flex-Foot Cheetahs'.

The Cheetahs were derivatives of Van Phillips' first creation back in 1982; they'd lost the heel plate and were now just a J-shape optimised in height and thickness for each athlete. The blades were tilted forward slightly by seven degrees so that the athlete now seemed to be standing on tiptoes. This would allow the blades to flex to the maximum, store huge amounts of energy and return as much of it as possible to the runner on push-off.

Pistorius' performances improved and he was selected to race for South Africa at the 2004 Athens Paralympics in the 100 and 200 metres T44, the race for lower limb amputees. He'd been running for less than a year and his start was terrible. His first race in Athens was the 200 metres; there were four false starts and, on the fifth, Pistorius stayed in the blocks expecting another. It didn't happen and by the time he started moving he was almost ten metres behind the leader, Brian Frasure of the USA. Pistorius threw everything into the race, chased Frasure down and stormed through to a new world record of 23.42 seconds. He ran better in the final, winning gold with another world record of 21.97 seconds.

Overnight, the 17-year-old was thrust into the limelight. He returned home a hero with the media loving his amputee-to-Paralympics story. Össur must have congratulated themselves on their choice of athlete. As an innovation company, they continued to improve their products and their blades had a shimmering black high-tech look that seemed to epitomise the age. And, with the impetuousness of youth, Oscar was fearless. His parents had always encouraged him to live life as a normal teenager. Why confine himself to the world of disability? If he was good enough, why not try for the Olympics as well as the Paralympics? The thought was compelling.

But with Össur suggesting the Cheetahs offered 'a powerful energy kick', some began to get nervous about their ability to enhance performance, not least the IAAF; they reluctantly agreed that he could run in able-bodied races until they could find out whether he had an advantage. We find ourselves back in rainy Sheffield in 2007.

The IAAF engaged Professor Gert-Peter Brüggemann of Cologne University to compare Pistorius with five able-bodied athletes. He found that Pistorius could swing his legs much faster than a non-disabled runner because his were so light and had such a low moment of inertia. Brüggemann's conclusions to the IAAF were clear: Pistorius expended 25 per cent less energy than an able-bodied runner. Pistorius was not just different, he was biomechanically off the charts. The IAAF quickly banned him from further able-bodied races.

Pistorius' management flew into a rage and assembled their own dream team of academics and compared him to a new set of athletes; they showed that he wasn't so different after all. However, they'd been a little too selective with the athletes they'd chosen (endurance athletes rather than sprinters) and sports scientists pointed out the flaws in their analysis.[124] Pistorius claimed that, yes, he might have an advantage towards the end of the race, but was disadvantaged coming out of the blocks. They took their results to the Court of Arbitration for Sport in Lausanne and, in May 2008, to many people's shock, Pistorius' complaint was upheld and he was cleared to run in able-bodied races.

The decision was so late in the season, however, that Pistorius didn't quite have the time to prepare and he didn't qualify for the Beijing Olympics. Instead, he went to the Paralympics and won three more gold medals. He spent the next four years training hard for both the London Olympics and Paralympics and in his biography claims he lost 17 kilograms: if true, this would represent over 20 per cent of his body mass. He qualified superbly for the London Olympics with a personal best in the 400 metres of 45.07 seconds, right up there with the finest sprinters in the world. He came second in his heat but crashed out in eighth place in the semi-finals.

With the Olympics out of the way, he turned his focus on to the Paralympics. There's a universe somewhere in which Pistorius competed, won medals and walked away a hero. But in this one, controversy was never far away and he was about to walk into a trap of his own making.

Growing pains

The suspicion that the Cheetahs might have given Pistorius an advantage over able-bodied athletes never really went away. In a newspaper poll of 40,000 people, more than 80 per cent were against him running in able-bodied races, as was the great 400-metre runner Michael Johnson.[125] It didn't help that Pistorius' own research team split apart over their divided opinions. One of the key researchers, Peter Weyand, was clear in his views: "From the instant we collected the gait mechanics data and saw how short his swing times are, we said to the group that it's really clear he's got an advantage."

His first event in the London Paralympics in 2012 was the 200 metres and he was the favourite to win. He set off well, taking an immediate three- to four-metre lead. But racing up the outside, in lane seven, was Alan Oliveira of Brazil, catching Pistorius with every stride. Oliveira passed him on the line, beating him by seven hundredths of a second to take the gold medal.

The crowd was stunned, the commentators were stunned, Pistorius was stunned. Even Oliveira looked stunned. This was an upset and Pistorius was not pleased. When he shook Oliveira's hand, he did a funny patronising bow of his head. His strange nod meant something, but what?

Pistorius believed that Oliveira had unfairly increased the length of his blades. It was just not possible to compete against his unusually long stride length, Pistorius said: Oliveira had an unfair advantage.

The irony was lost on Pistorius but not the rest of the world: while he'd maintained to the IAAF that he had no advantage over able-bodied runners, here he was accusing a rival in his own race of having an advantage over him by manipulating the design of the blades. Both couldn't be true. Pandora's box had been reopened.

Oliveira admitted that he had increased the length of his blades – by four centimetres – but claimed it was within the rules. The International Paralympic Committee agreed and said he could have gone a further four centimetres if he'd wanted to. They also pointed out that Pistorius could have raised his height by seven centimetres, but had chosen not to. The reason for the generous allowance in the length of the prosthetic was because it was unclear exactly what the height of a double amputee ought to be. There were anatomical tables that gave ratios of arms and trunk to legs for the general population which could be used as a guess, but an allowance was needed to allow for natural variations. This meant that athletes were able to play around with height to work out what worked best for them.

Oliveira certainly looked a little odd on the start line, as if he was on miniature stilts. Was Oliveira's stride length unusually long as Pistorius claimed? Or was his improvement more to do with hard work and training as Oliveira claimed and as Pistorius had always said about himself?

Making longer blades is not as simple as it sounds. A longer blade flexes more so is more prone to breaking. Lengthening the blades by two per cent as Oliveira did would have made them deflect around seven per cent more. The designers could have mitigated this by making the structure stiffer by making the square cross section deeper.[126] As with tennis rackets, the blades are made by laying up carbon fibre sheets in a mould, with the fibres aligned in different directions to give them both vertical and lateral stiffness. The designers would have tinkered with the angles of the sheets and the number of fibres per sheet – the more fibres, the stiffer the result. Since there are many combinations of lay-ups, the manufacturers would have optimised the blades to give the appropriate mechanical response and the right feel for Oliveira.

We can see where Oliveira's improvement in performance came from by comparing his performance in London to that in Beijing four years earlier. The first thing to notice in 2008 is that he used really short blades; the second thing is that he looked really young – he was only 16. He had short, fast steps and swung his legs in a sideways rotation rather than in a linear pumping action like Pistorius. His stride frequency was 4.3 per second and his stride length 1.9 metres: this gave him an average speed of 30 kilometres an hour. Pistorius' stride frequency in Beijing was similar to Oliveira's, but his stride length was a massive 2.2 metres. This meant he could run an average speed of 33 kilometres per hour, about ten per cent faster than Oliveira (who placed seventh).

Four years later in London, Oliveira looked distinctly taller. He was slow out of the blocks and two tenths of a second behind Pistorius by the first stride; by the ten-metre mark, he was four tenths of a second behind. Onto the final straight, Pistorius was in cruise mode and the commentators were confidently calling the victory his. But three metres behind, Oliveira had wound up his stride frequency to five steps per second, much faster than in Beijing. His stride length was still shorter than Pistorius' at only two metres, but his stride rate was so fast that he was running at 34 kilometres per hour compared to Pistorius' 33. It was inevitable that Oliveira would catch him and win.

Pistorius' claim about Oliveira was true, his stride length *had* gone up between Beijing and London, by about five per cent. But just as important, his step rate had also gone up, by an average of seven per cent. The combination of the two meant his speed had gone up by 12.4 per cent.

Between Beijing and London, Oliveira had changed his training, his coaches, his stride frequency, the height of his blades and his stride length. All contributed to his improved technique. Pistorius, on the other hand, was already at his peak and would have had to rely on everything going right on the day: it didn't. He broke the 200 metres world record in the heats with a time of 21.30 seconds. Had he run this in the final, he'd have beaten Oliveira by over a metre. The truth was that the world had caught up with Pistorius.[127]

The revolution in the design of racing wheelchairs and prosthetics occurred in tandem with the improvement in the manufacture of carbon fibres for use in planes. While sport is an early adopter, it's only in the last two decades that carbon fibre has been proven to be safe enough for planes such as the Boeing 787 Dreamliner: this is now 80 per cent composite.[128] Ironically, the burgeoning use of carbon fibre by the aerospace industry has meant that there is now sometimes a shortage for the sports industry.

A colleague of mine once told me how her young son was playing with his plastic action figures and feverishly hunting through his toy box. When she asked him what he was looking for, he replied that he was looking for the one with a missing leg. He wanted to play superheroes, he said, like at the Paralympics. But, sport seems to me to have moved on already from the materials revolution of the 1980s and 1990s. The next chapter describes the next revolution and how sport and culture are inextricably linked to our new definition of technology.

TWELVE

Sports technology in the new millennium

It was spring 2008 and I stood on a platform at St Pancras station, waiting for the Eurostar to Paris. My phone rang. "Hi," said the tinny voice. "It's Scott – I've got a proposal."

Scott Drawer was Head of Research and Innovation at UK Sport. It was unusual for someone to be ringing to give me something; I was usually the one chasing after them. But Scott was a creative thinker and had a different way of doing things. His job was to source technological solutions to the problems facing our Olympic teams, to provide them with the best bikes, boats, shoes, wheelchairs and so on.

"I've got some innovation funding. I'd like you to be one of our innovation partners." He didn't mean me personally – I can't do anything practical any more – he meant my team.

"The deal is this – whatever funds I give you, you have to match it pound for pound. Are you in?"

I didn't hesitate. "Of course we're in," I shouted above the roar of the trains. I had no idea how I was going to match it: if it was a million pounds, I was in big trouble. In my excitement, I'd forgotten to ask him what he actually wanted us to do, but no doubt we'd work that out later. It seemed that we'd just become a UK Sport Innovation Partner. It felt good.

Back in 2008, when we got the partnership, we'd mostly been involved in structural aerodynamics, design and testing. What I was about to find out was that our Olympic sports didn't want that from us any more. That was for the big boys, the likes of BAE Systems, McLaren and Frazer-Nash. But it seemed that most of the coaches had read *Moneyball*, the best-selling book by Michael Lewis[129] and what they now wanted was analytics.

If you don't know the *Moneyball* story, it's about how the cash-strapped Oakland Athletics baseball team from California matched the spending power of the bigger and richer baseball teams. The hero of the story was Billy Beane, the general manager of the Oakland A's. Baseball was run by those who made decisions through intuition, gut instinct and experience, those with the arrogance to stand in a room of tobacco-hawking men and stare them down.

Beane, however, had been introduced to the research of a baseball fanatic called Bill James who had manually collected baseball data and analysed it. He had one question: What does it take to win? He created an equation that predicted how many runs a team would get using just two pieces of information for each player. It was remarkably accurate. He quickly showed that the coaching dogma spouted by the baseball fraternity was all wrong.

Of course, he was dismissed by the power brokers because he threatened their control over the game. How could a stats nerd like James, who'd never played baseball in his life, know more than the seasoned pros who'd been in the game since forever? Despite this, Beane employed a Harvard graduate to expand on James' research and help him assemble a new team based on these radical ideas. The A's improved dramatically and went on to get the longest continuous series of wins of all time, 20 wins in a row. Beane rejected a $12.5 million job offer from the Boston Red Sox and his fame was secured when Brad Pitt played him in the 2011 movie of the book.

The *Moneyball* effect was swift and the rest of the sporting world woke up to the new world of analytics. My dictionary tells me that analytics is 'the systematic computational analysis of data or statistics'; it might be new to sport, but people have been doing it for centuries.

Take Florence Nightingale, for example. You may have heard of her – she was the nurse who looked after soldiers in the Crimean War in the 1850s. The classic image of her is as 'the Lady with the Lamp', wandering around darkened wards in the middle of the night, caring for frightened soldiers near to death. What you might not know is that she was also brilliant at analytics.

She collected data on the causes of death of her soldiers and showed the government that very few of them actually died of their wounds: most of them died because of the unsanitary conditions of the hospital itself. She presented the data as a beautiful pie chart so that it was easily digestible by the politicians and suggested they put in better sanitation. It worked. She then used the same approach over the next couple of decades to lobby for better sanitation and conditions back in Britain. The Public Health laws she promoted increased the lifespan of the average Briton by around 20 years.

Analysis is something scientists do every day of their lives, but what Florence Nightingale realised was that not everyone likes numbers. If you want to convince someone of a course of action, then you need to make the numbers palatable. Her pie charts were some of the earliest examples of what we now call infographics, one of the key components of modern analytics.

Whether it's politicians, coaches or football managers, the approach is the same. Collect data, analyse it, present it in a meaningful format. Then comes the 'so what?' The point of analytics is that the evidence should be used to guide decisions but, quite often, evidence goes out of the window when those in power come to make them. Football managers and prime ministers are not so different.

One of Scott Drawer's favourite quotes was from a Dutch business theorist called Arie de Geus who said, "The ability to learn faster than your competitors may be your only sustainable competitive advantage."

The UK Sport partnership would allow us to take de Geus'

words and put them into practice. It would take us from Beijing 2008 to London 2012, then to Rio 2016 and now on to Tokyo 2020. Our work would help our Olympic teams to learn faster and understand more than our competitors. We had just entered the world of *Moneyball* and analytics, and first, we had to decide what data to collect.

We didn't get a million pounds from Scott and our budget had significantly fewer zeros, but this became an advantage as it focused the mind. There were only two criteria for the projects we would do: first, they had to be value for money; second, they had to improve medal chances. Nothing else mattered. I remember one high-profile sport being taken aback when we declined to do their project; they might have been a major Olympic sport, but it would have wiped out our budget in one go.

We sought out champions in the sports who were already doing their own analytics. Coaches, performance analysts and even athletes were collecting data wherever they could and putting it manually into spreadsheets scattered far and wide across the country in hundreds of different formats. Spreadsheets were everywhere and used for everything. Performance analysts had become adept at using spreadsheets as an analytical tool, but the sheer amount of data was becoming unmanageable. Our collective plan was to help them with the data collection and follow Florence Nightingale's approach to make it palatable to the coaches and athletes.

I remember a diving coach called Adam Sotheran coming to us somewhere around 2008, brandishing his new iPhone by the side of the pool; he was very excited. "I want to collect videos with this. And attach data to It. And tag the videos with data and send it to my divers' phones. And…"

Adam's ideas predicted what others would want: to collect data and videos out in the field, store it, send it back, share it and make sense of it all using nothing more than a mobile phone (tablets and iPads were still things of the future). Simon Goodwill stepped up once again, rapidly retrained himself as a software developer and started to create the systems the Olympic teams wanted.

The question was always the same. If the team introduced something new, whether it was a coaching technique, a bike, boot or blade, how could they tell if it had improved performance? This needed data but the field was their laboratory, exactly as I'd found back in the 1980s. Just as in baseball, there was reticence in some quarters – how could stats replace the wisdom of years of coaching? But there were enough champions in the system: Rob Gibson at GB Boxing; Becky Edginton at gymnastics; Julia Wells at canoeing. They knew that data alone didn't provide the answers, it also needed their expertise to translate it into the sporting intelligence that could be passed to the athletes. The systems weren't replacing the coaches, they were enhancing them.

Simon suddenly had more work than he could cope with and Scott's budget for us began to grow; we worried how we might cope with the workload. Then one day, Chris Hudson walked into my office and said, uncertainly: "I'm a software engineer. I'm bored with my job, but I love sport. I don't suppose you need a programmer, do you?"

I thanked the gods of serendipity, hired him on the spot and Chris and Simon have been the mainstay of the UK Sport work ever since.[130] We started to collect and store data. There was information on the athlete such as height, weight, lengths of their arms and legs, diameters of their muscles. Injury data – due to its sensitivity – went somewhere central locked behind several layers of security. Then there was performance data such as scores, distances and times.

It's all in the timing

The clichéd image of a coach with a stopwatch hanging around their neck exists for a reason. Time has been used to measure sports performance since, well, the beginning of time. The stopwatch wasn't designed with coaches in mind, it was actually intended to allow doctors to measure the pulse. Sport appropriated the stopwatch in the 1690s to help satisfy our insatiable appetite for gambling. Pedestrianism was a cross between running and walking and was popular in the 18th century. People would challenge each other to race long distances or a particular distance in a set time, ten miles in an hour, say. Timing the hour accurately was essential for both sides of the bet.

Early clockwork stopwatches had an accuracy of one fifth of a second. The invention of the transistor in the late 1940s created the age of transistor radios, calculators and computers. But most importantly for sport, it heralded the blinking red digital clock, first used in ski racing and with a claimed accuracy of one thousandth of a second. A paper by a US Olympic Committee researcher at my second sports engineering conference, however, showed that the internal electronics of any system were often only good enough to allow a hundredth of a second, no matter what the claim was.[131] The first digital watch appeared in 1971 and, by 1975, fully automated digital timing was introduced into sport, producing the two tenths of a second upward jump in running times we saw in Chapter 1.

Putting your heart into it

Performance in most sports improved rapidly due to better coaching, nutrition and sports facilities. Sports science became an established discipline in universities and experiments in laboratories began to reveal how performance could be improved. Often, heart rate was used to monitor these improvements, although it wasn't particularly easy to do. The gold-standard method was relatively cumbersome and used 12 electrodes to record electrocardiograms (ECGs); pairs of the electrodes measured the waves of electrical signals given out by the heart as it pulsed. Unfortunately, everything was wired so that it was only really suitable for laboratory tests.

A second, simpler system took a single measurement of the pulse by shining a green light directly on the skin and used a detector to pick up the reflected light: the more blood that rushed past, the less light was reflected. The best place to measure was either the earlobe or the end of the finger but even this simpler system still required wires. The appearance of the digital watch had got people thinking. One of those was Seppo Säynäjäkangas from the electronics department of the University of Oulu in Finland. Out cross-country skiing near his home in Kempele, he met an old friend working as a coach. Wouldn't it be great to measure heart rate while out skiing? Wouldn't it be great to see it on a watch?

His first portable solution used an optical sensor system on the finger but then he opted for the ECG approach, launching a version in 1982 that had two electrodes embedded in a strap fitted tightly across the chest. He called it the Sport Tester PE2000 and set up a company called Polar to sell it.

Other companies copied and introduced their own systems. Some didn't work very well and even the best ones measured heart rate around five to ten beats lower than the 12-electrode ECG.[132] But what did it matter? Athletes had never been able to measure heart rate during exercise before so didn't really know what to expect. As long as the watch didn't give out crazy answers and heart rate went up during exercise and down during rest, the absolute number didn't really matter.

Paula Radcliffe, one of the greatest endurance runners ever, was an advocate of the heart rate monitor and used it to gauge performance while out running. Her coaches actually wanted to measure blood lactate, but this was too difficult out on the road: heart rate would be their proxy. The level of lactate is an important indicator of how the body is responding to exercise. At slow speeds, the body uses energy from oxygen coming in through the lungs. As the speed increases, however, the body can't process the oxygen fast enough, so the body looks for an alternative energy source. Lactate in the blood fills in for the missing energy a bit like an emergency battery but, similarly, doesn't last very long. As the lactate is converted to energy, acidity in the muscles rises which reduces their efficiency and leads to fatigue. Eventually the muscles start to burn, which is blamed on 'lactic acid';

the worst-case scenario is that you grind agonisingly to a halt.[133]

A few years back, I was in training for a marathon and read how Paula Radcliffe had used heart rate to predict lactate; I decided to give it a go.[134]

A physiologist at the university – Alan Ruddock – agreed to do the same test on me that Paula had gone through. The plan was to put me on a treadmill, crank up the speed in two-minute bursts and test my blood as we went. We started with a gentle ten kilometres per hour jog – this equated to a marathon time of about four hours and 13 minutes. We would end at 16 kilometres per hour, equivalent to a marathon of two hours and 40 minutes. I was certain this was beyond me.

I climbed on the treadmill and, disconcertingly, Alan strapped on a rope hanging down from a frame above my head. Health and safety gone mad, I thought. I clipped a Polar heart rate strap across my chest and made sure it was tight: Alan suggested I use a bit of saliva to dampen the electrodes to get a good connection with the skin. I checked my resting heart rate on the watch strapped to the frame of the treadmill – like my age, it was just over 50.

Alan's test was designed to find the speed and heart rate where my body would begin to rely on lactate. He would measure it by pricking my finger at each speed with a small disposable pin and putting it in an automated test device. Lactate is measured in thousandths of a 'mole' per litre of blood (a mole is a scientific standard of 0.6 million million million million molecules). Alan pricked my finger with a pin and put it in a machine: my lactate was one thousandth of a mole per litre. I started jogging and my heart rate jumped immediately to 120 beats per minute. Alan pricked my finger again: still one thousandth of a mole per litre. Alan pushed the speed to 11, 12, then 13 kilometres per hour, measuring my lactate as we went. My heart rate rose to over 150 beats per minute but my lactate remained at the same low level. My breathing became laboured at 14 kilometres per hour, with a bit of a 'hah' on every other step. My heart rate increased to 155 per minute and my lactate doubled. The 'hah' became a heavy 'huh' at

15 kilometres per hour and, with my heart rate at 170 beats per minute, my lactate was now four thousandths of a mole per litre.

"Can you go any faster?" he asked.

I gasped a yes and grimaced as the treadmill sped up. The two minutes at 16 kilometres an hour felt like a lifetime. My breathing was a heavy 'huh, huh, huh' on every step and I almost fell off the treadmill when Alan called time: the rope saved me. My heart rate had peaked at over 180 beats per minute and my lactate was a burning six thousandths of a mole per litre – six times the starting value.

As I recovered on a chair, towel on my head and my cheeks glowing red, Alan explained the data. I was really quite suited for long distance running, he said. He didn't actually say I was rubbish at sprinting, but that's what he meant. My lactate profile had stayed low until 14 kilometres per hour after which it had shot up rapidly. My heart rate at this threshold was about 160 beats per minute. This was the key number, he said. Keeping my heart rate lower than this for as long as possible would keep me cruising at about 13 kilometres per hour without going into the rapid fatigue that lactate would cause. If I kept up the training, he estimated that my best marathon time would be three hours 15 minutes.

I compared my lactate profile to Paula's. When she was 18, her lactate threshold was at 15 kilometres per hour. But then she started training properly and, over the next decade, her body adapted to the lactate punishment to push her threshold up to 19 kilometres per hour. Paula Radcliffe got the women's world record in the London Marathon of 2003: her time was two hours 15 minutes and 25 seconds. This gave her an average speed of 18.75 kilometres per hour, as close to her lactate threshold as you could probably get.

The trend for heart rate monitors in recent years has reverted back to the optical type to reduce the reliance on a strap around the chest; it doesn't measure in the earlobe or on the end of the finger but at the wrist. A company called Mio pioneered the development of a sensor that used two low-energy, light-emitting diodes pressed against the flesh on the underside of a watch. A photo-sensitive chip a few millimetres across nestles between them, picking up the pulses of reflected light. A microchip in the watch counts the pulses, filters out the noise and presents the heart rate on the watch's screen.

I bought a TomTom watch which had one of these heart rate sensors incorporated into it and put Alan's running advice into practice at the 2015 Manchester Marathon. I set out at 13 kilometres per hour as he'd suggested and kept my heart rate below 160. My heart rate crept up gradually from about halfway and, with five kilometres to go, it was at 170 per minute, well over my lactate threshold. My legs started to get heavy and my speed began to drop; it was just a matter of hanging on to the bitter end.

I turned a corner, saw the finishing line and found the last vestiges of energy to sprint. Suddenly, it was all over and I looked at my watch. Three hours 15 minutes and 19 seconds, an hour behind Paula Radcliffe's world record but just as Alan had predicted. The trick with the heart rate monitor had worked.

(Confession: a year later, the BBC carried an article with the following title: 'Greater Manchester Marathon course was 380m short, says measuring body'.[135] It doesn't sound much, but it would have taken me another two agonising minutes to run that extra distance, pushing my time up to three hours 17 minutes. *Darn you Manchester Marathon*, I'll have to do it all again.)

One step at a time

In the world of sports equipment, there is an unwritten law that says technology should improve every year. Like washing powder, the next version must always be 'new and improved', hoping you'll forget that last year's version was already the ultimate product. Wearables are fuelled by the desire for data, so manufacturers of sports watches in particular are keen to pack in as many sensors as possible, all able to give you feedback on a different aspect of your performance.

Probably the most common sensor is the accelerometer. In my physics class, back in the early 1980s, one of my assignments was to design and test a device that measured something in the real world. I don't know why, but I chose to make an earthquake detector. I suspended a metal plate horizontally on springs parallel to another one attached rigidly to a bench above it. I applied a low voltage between the two to charge them up so they would act as a large capacitor. The idea was that when an earthquake came – and I really hoped one would – the suspended plate would vibrate up and down, changing the capacitance. I would measure the changing capacitance which would mimic the shaking of the earth.

What I didn't realise was that I'd created a crude accelerometer. Almost at the same time, about 5,000 miles west and close to the San Andreas Fault, Stanford University engineers were creating the world's first miniature accelerometer. This used exactly the same principle as my crude creation, in that it measured the change in capacitance between two moving parts. The main difference between their device and mine was about two kilograms of weight. Their accelerometer was only a few millimetres across and elegantly shaped like a set of flat intertwining fingers. The Stanford accelerometer took another 15 years to go into production; it didn't measure earthquakes, however, but was used to detect collisions for car airbags. The cost of accelerometers dropped and they were incorporated into games controllers such as the Nintendo Wii

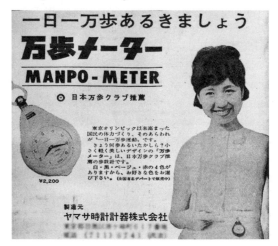

Figure 41. The original 1960s Manpo-meter, used to encourage the walking of 10,000 steps per day in Japan.

where they were used to deconstruct hand movements into the on-screen antics of avatars waving wands, axes and tennis rackets.

In sport, the first thing the accelerometer was used for was to replace the humble pedometer, a relatively basic device used to count steps. These had become popular in 1960s Japan when a young professor at Kyushu University called Dr Yoshiro Hatano had promoted them as a device to encourage physical activity. He was worried that the gradual westernisation of the country after the Second World War would lead to an increase in calorie intake and an obesity epidemic. He'd worked out that a person walked on average about 5,000 steps per day and suggested that increasing the number of steps to, say, 10,000 per day, should keep obesity at bay. Dr Hatano endorsed a mechanical pedometer device made by a watchmaker and they called it the *man po kei*, Japanese for '10,000-step measure'. It became so popular that any device used to count steps in Japan is still eponymously called a *Manpo-kei* much like 'Hoover' represents vacuum cleaners and 'Google' represents search engines.

Early devices were relatively simple and used a lever that would bounce up and down during walking to rotate a simple analogue dial. The accelerometer that replaced it was much more robust, small enough to fit in a watch and didn't have the annoying clicking sound made by the levers. Another advantage was that the accelerometer signal contained much more information: not only could it count steps, it could also give the step frequency and the intensity of the activity. Walking, running and cycling all gave different acceleration patterns which could be automatically detected, and it could even tell if you'd given up completely and taken the car.

Accelerometers are one of a new suite of micro-electronic measurement systems, called MEMs for short, and they've inevitably made their way into sports equipment such as footballs, tennis rackets and ski boots. adidas put a Sputnik-like module inside their smart football that has three accelerometers and three gyroscopes to give linear and rotational measurements in all dimensions. It has a rechargeable lithium-ion battery, microprocessor, digital compass and wireless connectivity using a standard chip called Bluetooth

Low Energy. All of this sits within a sphere a little smaller than a golf ball and is held at the centre of the inflated ball by eight struts made of Kevlar.

Figure 42. The adidas smart ball containing a three-axis accelerometer and gyroscope, digital compass and rechargeable battery. © *adidas*.

When you take a free kick with the ball, it gives you speed, spin and a simple plot of the trajectory. This is surprising since it works primarily with accelerometers which are best at measuring acceleration, not distance or speed. It manages this using a mathematical process called integration. Integrating acceleration gives speed, integrating speed gives distance. The other sensors help with working out the trajectory: the compass gives the direction of the kick while the internal gyroscopes give the rate of rotation of the ball. Even this is still not enough to get sensible answers out of the ball and the instructions tell the kicker to place the ball in a particular orientation and identify which kicking foot is being used. This tells the Sputnik-like sensor inside which way is up and which way the ball is likely to rotate: a kick with the right foot tends to make the ball rotate anti-clockwise when looking from above and the ball veers off to the left.

Kick the ball, and the speed, spin and ball trajectory pops up on your phone, ready for sharing on the social media platform of your choice.

Where are you?

Sensors are now everywhere: in smartphones, in watches, even in shoes. The Nike+ was a little pod about the size of a thumb that sat inside the sole of your shoe; it had a piezoelectric accelerometer that measured steps like a pedometer and used an estimate of stride length to give distance travelled. All pedometers calculate distance this way but can be wildly out if the stride length estimate is wrong. A more accurate way to measure distance is by using global positioning satellites, or GPS for short. The first satellites were launched by the American Air Force in the late 1970s but it took another two decades for a 24-satellite system to become fully operational. GPS generally refers to the American system, but the Russians have one too called GLONASS for *Global'naya Navigatsionnaya Sputnikovaya Sistema*. It's very satisfying to say out loud: *Global'naya Navigatsionnaya Sputnikovaya Sistema*. The systems need to be as accurate as possible and the satellites use atomic clocks synchronised to an accuracy of 14 nanoseconds (to give you an idea how short this is, light travels only three metres during it).

The satellites move around carefully spaced orbits about 20,000 kilometres above the earth with ground bases continually checking where they are. The synchronised satellites continually broadcast their time and position and, the further away the satellite, the longer the signal takes to get to you. My watch receives the signals from each satellite and uses the travel times to work out their distance, needing to see at least four to get an accurate fix on my position. With four or more signals, my watch can use trilateration (the 3D equivalent of triangulation) to work out where I am to an accuracy of about five to ten metres.[136]

Testing by sports scientists showed that early GPS units weren't quite good enough for short distance sports, particularly if sprinting was involved. One of the reasons for this was that the GPS units only took measurements once every second, limited by a lack of memory and low battery life. If a 20-metre sprint takes five seconds, then the watch will only have recorded five data points, each about four metres apart. If you plotted these points on a graph, every point would have superimposed on top of it the five to ten metre inaccuracy from the satellites: this makes short distance measurements unfeasible. Despite this, there were still hundreds of research projects on movement patterns in soccer, Australian football, rugby and hockey.[137]

When I go out for a run and switch on my watch, there is usually a delay while it searches for the satellites. Runners with GPS on their watches adopt a funny position at the beginning of a run as they seeks out the satellites, arm stretched out horizontally like an aerial, face pointed beseechingly to the sky as if it's possible to actually see the satellites from 20,000 kilometres away. If it's before a race, there is usually a frown on the face and a look of panic as the countdown begins.

Although my watch has a GPS receiver, triaxial accelerometer, gyroscope and an optical heart rate monitor, it just looks like an ordinary black digital watch. Some also have a barometer to give altitude, a temperature sensor and, of course, the all-important music player with Bluetooth headphones. I now get to record every run or ride using my watch and upload it to a website to share with my friends.

I love my GPS watch and it's as important to me as my phone. But, I can feel a weird psychological thing going on. If I do a run and don't record it, the run may as well not have existed. I get really cross if the battery runs out halfway through because I'll never know how good that run really was. Apparently, at least 40 per cent of people suffer from nomophobia ('no-mobile-phobia'), the fear of being without your phone. I think I have 'nochronophobia' (I made the name up), the fear of being without my sports watch.

Thirsty for knowledge

A few years ago, I was talking to an academic at an analytics conference about the explosion in analytics. "We're thirsty for knowledge, but drowning in data," he said. It was a neat way of making a point: there are now a multitude of ways to collect data, but on its own it isn't enough to give you answers. You have to analyse it for a purpose, just as Florence Nightingale did.

Another sport that has embraced baseball's *Moneyball* approach is soccer. If you want to know 'why everything you know about football is wrong', read *The Numbers Game* by Chris Anderson and David Sally. This does for soccer what *Moneyball* did for baseball and strips away the entrenched beliefs of managers and coaches using data and statistics. For instance, while a single goal is worth about a point per match, stopping one is worth a whopping 2.5 points per match. Score goals by all means, but don't let any in.

You wouldn't expect mathematical models like those of great mathematicians like Poisson and Bernoulli to also explain something as complex as human behaviour. Yet surprisingly, Anderson and Sally show that they do. One model I've found, and which intrigues me, is Benford's law. The law's discovery actually goes back to an American astronomer in 1881 called Simon Newcomb. He used logarithm tables to make calculations and noticed that earlier pages where the digit one was looked up were much more worn than later twos, threes, fours and so on. He wrote down a simple equation to explain it using logarithms (naturally). About 50 years later, a physicist called Frank Benford rediscovered the law and found that it applied to lists of numbers such as voting patterns, the stock market and even the incidence of lightning strikes. The law says that digit one will always lead 30 per cent of the time, digit two 18 per cent of the time, digit three 13 per cent of the time and so on.

I looked at the number of passes made in soccer using data taken from Anderson and Sally's book. If you compare Benford's law to the data for the 2011–12 Premier League season, the ball is intercepted 34 per cent of the time on the first pass, 18 per cent of the time on the second pass and so on. It's remarkably close to Benford's law.

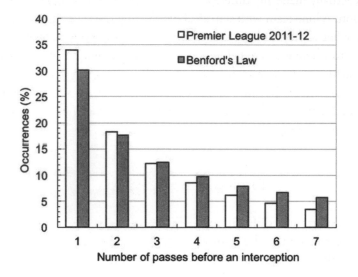

Figure 43. The number of the pass on which an interception occurred in the 2011–12 Premier League compared to Benford's law of the first digit.[138]

What does this tell us? Benford's law predicts what will happen on average (although my wife says it actually proves that soccer is pointless). Thus, if the passing profile for a particular team is different to the average, then the coaching methods are probably making a difference. If the plan is to keep hold of the ball and control the game, for instance, then the distribution would be skewed to a higher number of passes. If the plan is for a long ball game where the ball is punted forwards and more likely to be intercepted, then the distribution would be skewed to a lower number of passes. Mathematical laws such as Benford's law can be used to work out if there is anything going on that is different to the norm.

I'm not privy to the backroom data of big clubs like Manchester United, Real Madrid or Barcelona, but this is just the sort of nugget of understanding the analysts yearn for. And now there are a multitude of companies to help them, such as Opta, Prozone, Infostrada and StatDNA, not just for soccer but for all sports.

The one statistic I would like to know is whether the use of data has actually made any difference to league position at the end of the season. A lone team with statistical insight might have an advantage, but what if they all have the same statistical knowledge? Aren't we back to where we were before – everyone being equal? If there is an imbalance, then one team will have an advantage over the other, otherwise they will tend to cancel each other out. Anderson and Sally showed that soccer, like many team sports, is a game of equilibrium. This is something I'll return to in the final chapter.

Turning data into gold

Team sports are difficult to analyse because of the amount of variables involved. The number of players, the team selection, the opposition, the weather, the referee's decisions: all can affect the outcome. It's easier when there is just one person involved, a single athlete with a single coach and just one task: running the 100 metres; throwing the javelin; performing a vault.

After a decade of work as a UK Sport Innovation Partner (now under the auspices of the English Institute of Sport), my team has created over 100 different systems to collect data on the UK's Olympic athletes. The simplest collect just video and store them in a safe and convenient place. One of the first was gymnastics software for British Gymnastics called iGym which was installed in their elite gymnasium in Lilleshall. Cameras on the wall automatically monitored the different apparatus. In the vault discipline, a gymnast runs down a track, jumps onto a springboard and catapults themselves over a pommel horse, performing fantastic twists and somersaults before landing on a thick blue pad on the other side.

Simon Goodwill mounted a camera on the wall to continuously record the view on its internal memory at 25 frames per second.

The coaches didn't have time to continually start and stop the video so he created an image processing algorithm to detect the gymnast flying past which automatically stored the video, giving it a filename and a timestamp. This allowed the coaches to continue coaching and the performance analysts to retrieve the videos they wanted. They found the occasional surprise in them as some of the younger gymnasts had figured out how the triggering worked. They would climb on the pommel horse and, rather than a video of a gymnast flying past, all the analysts would find on playback was a cheeky young gymnast jumping up and down waving at them.

One of the next systems was iDive, developed with coach Adam Sotheran in the international diving pool in Sheffield. A camera was set to record dives from the three-metre springboard. The athlete would take two bounding steps down the board before landing on the end with both feet; the board would then catapult them high into the air, allowing them to complete a complex gymnastic manoeuvre before pointing themselves downwards towards the water's surface. The idea was to enter the water with as little splash as possible.

The diver would reappear at the side of the pool to look up at Adam, who would point to the huge screen behind him which was now automatically playing the dive in high definition slow motion. If he wanted, he could use his iPhone to control the video and highlight the key points he wanted to get across. As the diver climbed out to do a repeat dive, Adam might save the video if it was good enough and send it to the diver's phone so they could talk about it later in the café. He might even tag the video with text and body angle measurements as take-home messages, just as he'd wanted.

The boxing system was unsurprisingly called iBoxer and is now one of the most sophisticated performance analysis systems in the world. GB Boxing built a beautiful boxing gym at the English Institute of Sport in Sheffield, inheriting and installing the ring from London 2012. They put cameras over the five boxing rings to record their own training bouts as well as those of visiting teams training there. Rob Gibson, the performance analyst, began the laborious task of importing the results of every fight from every tournament into the database. Now, if they want intelligence on any boxer, it is there, coupled to videos of the fights. The names of the judges and referees are noted in case the fighter has to overcome any perceived 'subjective biases' that might exist (the boxing coaches put it in more colourful language). The gym with its coaches, performance analysts, support staff and facilities is now one of the best in the world: no wonder Anthony Joshua, the current heavyweight champion of the world, still trains there.

There are now many of our performance analysis systems scattered across the UK: iTaekwondo in Manchester; NEMO for swimming in Loughborough; CanoeSPI at Lee Valley. These systems helped Team GB's athletes get 24 medals in London and 42 medals in Rio. We're now working towards Tokyo 2020. What do they do with all the data? Ultimately, it ends up at UK Sport headquarters where they have the Sports Intelligence Unit. The athlete is at the centre of everything: analysts make predictions on how they will fare at the next Olympic Games, and the coaches and support teams use data to enhance performance. If it all goes to plan, the Medal Tracker Board at UK Sport will show that plenty of gold medals are on the way.[139]

Whenever I talk about sports technology these days, most people think about wearables and smartphones. But look into the past and the dominant technology of the time was whatever was available: 20 years ago it was carbon fibre, 100 years ago it was wood. What will happen as we move into the near future? What technologies will influence sport as we go a few centuries further?

This is the subject of the final chapter.

THIRTEEN

The next big thing

The video showed a skier slowly carving turns down the hill and ploughing to a halt at the bottom. Suddenly, a gigantic pair of godlike hands came down from the sky and the skier was whisked into the air, dangling limply upside down. The skier was a miniature robot; the hands belonged to its Japanese creator. I was in a presentation about the mathematics of ski robots at one of my earliest conferences on sports engineering and it showed me two things: one, the Japanese love robots; two, we will race *anything*.

The simple bet is at the heart of sport in the sense of 'I bet I can beat you to that tree', or, 'I bet this robot skier can beat yours'. This is all sport is, a set of arbitrary wagers that have been codified over the years. And some of the sports we've created are really bizarre. Who has ever thought that golf, the modern pentathlon or synchronised swimming ever made sense?

Humans are almost unique on the planet in their love for technology and if it can help us win our sporting bet then so much the better. That was why Coburn Haskell wanted his rubber-cored golf ball, Christine Nesbitt her clap skates and Ilie Nastase his spaghetti racket.

So, what's the next big thing?

Materials are the foundation for much of sports technology. The ancient Greeks used stone and sinew for their starting gates, wood and leather for their javelins, lead and bronze for their jumping weights. The Victorians used pigs' bladders and then rubber for their footballs and more recently we have used carbon fibre for almost any sporting application we could find. New materials have been the catalyst for rapid change in sport, so which material is next?

Graphene looks like the bookies' favourite. It won a Nobel Prize for its Manchester researchers in 2003 and has been hailed as a supermaterial to change the world. Graphene is a single atom hexagonal lattice of graphite that's 200 times stronger than steel, more conductive than copper and thinner than paper. The problem at the moment is that no one has managed to create a sheet of it more than a few millimetres across. But, that hasn't stopped people like Head incorporating it into tennis rackets by mixing it with other materials. The claim is that it makes the equipment lighter and easier to manoeuvre.

This has hardly set the sporting world alight. Players don't necessarily want a massless tennis racket; take away the mass of the racket and you might as well go back to using just your hand. The opportunity graphene gives, however, is to play around with the racket's mass distribution and moment of inertia just as the ancient Greeks taught us,[140] and as we saw in tennis racket development of the 1980s.

Callaway, never shy in the technology department, have golf balls with graphene-infused polybutadiene wrapped around a soft inner core. The softer the core, the greater the deflection during impact with the club and the greater the energy stored. This energy is released as the ball returns to shape at the end of impact, giving it a higher coefficient of restitution. The graphene is there to stop the soft core splitting apart, and Callaway claim that the ball is a game-changer. As with tennis rackets, I'm not sure that the change will be all that big, mostly because the rules of the R&A and the USGA won't allow it to be.

I think it's more likely that graphene's main advantage will be its ability to conduct electricity. Graphene in the face of a club or the frame of a racket might enhance the strength, mass distribution and feel, but could also transmit data from inbuilt sensors. On the human body, data is currently collected using wearable sensors but manufacturers are desperate for a good way to embed them into textiles and clothing. This requires the textile to be flexible, strong and carry electrical currents through it. Once graphene's manufacturing issues have been solved, it could satisfy all these conditions; it could even provide protection from impact. This is just one of the reasons that different countries are spending millions trying to find the secret to the manufacture of graphene. The one who gets there first will make millions in sport and billions in everything else.

Mind over matter

Golfer Ben Curtis entered the 2003 Open Championship as a 300 to 1 outsider. It was his first major championship. Surprisingly, he won by one shot and jumped a few hundred places in the rankings during a single weekend. A few years later, a group of researchers from Virginia, Tubingen and Utah took Curtis' Titleist putter and gave it to 20 golfers to see how they would perform, comparing them to a control group with a similar putter.[141] Those with Curtis' club holed on average 5.3 times out of ten compared to 3.8 times for those without. What was it about Curtis' putter that made it so much better? Was it the design? Was it the material?

The truth was that the two putters were identical. The only added ingredient was that the golfers using Curtis' putter *thought* it had belonged to him because the researchers had told them so. One of the reasons they'd performed better was revealed by asking the golfers to estimate the size of the hole before the test: the group with the Curtis putter estimated the area of the hole to be 21 per cent bigger than the control group. Just believing that the club had belonged to Ben Curtis was enough for the players to believe the hole was bigger and easier to hit into.

The connection between a technology and our faith in its performance-enhancing ability is often enough to make it happen. The golf researchers called the effect 'contagion', where the mere fact that a past master had used the club imbued it with qualities the next user would inherit. The fact that *gloios*, the mix of sweat and oil the ancient Greeks sold as an embrocation, had been scraped off an athlete hero means that it might just have worked.

The catch-all name for the contagion effect is *placebo*. In Latin, this means 'I shall be pleasing'. Placebos have been shown to have real physiological effects: heart rate, blood pressure and respiration have all changed through their use.[142]

Placebo studies in sport have mostly looked at supplements, steroids and caffeine to see if their effect is psychological rather than physiological. Christopher Beedie from Canterbury in the UK gave placebos to cyclists but told them they were caffeine supplements. Their power output improved by about three per cent, similar to that seen in studies on real caffeine. In another experiment, runners were given a drink described as 'super-oxygenated water', and their performance in a five-kilometre run improved by over a minute, even though the substance doesn't actually exist. The least talented runners improved by over two minutes.[143] Telling someone about the scientific efficacy of a sports product seems to produce a performance improvement, whether there's a real effect or not. The mere expectation of a better performance seems enough for many people (although not all) so that, in sport, even non-existent things can be made to work.

This points, perhaps, to one of the next big things in sports science – the brain. The brain and the central nervous system are made up of neurons linked together through neural networks that transmit electrical signals. The signals can be externally influenced by electrical or magnetic fields. The practice of electro-convulsive therapy in the early 20th century is a stark reminder of that. The part of the brain that controls movement is called the motor cortex, and sits towards the front of the skull. If we can interact with the electrical signals there, then we may be able to enhance movement and improve performance.

A San Francisco company called Halo has produced a headset that looks like a stylish set of headphones, but actually uses the method of transcranial direct current stimulation (tDCS) to stimulate the motor cortex. It puts a small direct current across the surface of the head between two connectors; the positive one is called the anode, the negative one the cathode. The electrical potential in the neurons near the anode is raised so that when an external stimulus comes along, the neurons are primed to receive it. The idea is that this makes learning new tasks easier.

When Arie de Geus told us to learn faster than our competitors, he probably didn't have tDCS in mind, but Halo did, and they claim that their device improves strength, endurance and muscle memory. Already, players for the Golden State Warriors basketball team have given it a go, although there are no public comments so far.[144] The United States Ski and Snowboard Association also tried it during their preparations for the Pyeongchang Winter Olympics. While the US Olympic team had their worst performance for 20 years, the snowboard team won four of their nine gold medals and got seven medals in total.

Does it really work? Ten cyclists were tested with a tDCS system (but not the Halo).[145] The cyclists' power outputs increased by around four per cent compared to the system when it was switched off. Since the researchers didn't measure the cyclists' baseline performance without any sort of headset, we don't know how much the placebo effect improved performance on its own, but previous studies have shown that this is likely to be a few per cent. tDCS might well have had an overall physiological effect, but some of it was likely to be placebo.

This is great news for products like Halo. Tentative data shows that 20 minutes of tDCS prior to exercise might increase performance, at least temporarily.[146] Even if the electrical stimulation doesn't actually work for the user for some reason, the placebo effect means that they will probably believe it does, and this might be enough for them to improve anyway. Products like the Halo are a neat solution to the placebo problem: while evidence shows that the placebo effect really does work, people are unlikely to pay for it. But they will pay for the performance-enhancing effect of tDCS, and they get the placebo effect thrown in for free.

It's likely that more products will look to improve performance using the brain. Sleep analysis is an area that is expanding rapidly, and the data collection systems my team have created for Team GB often have questionnaires asking about the quality and quantity of sleep. But what can you actually do if you sleep badly and this affects your performance? Products such as mindGear use the same tDCS technique as Halo and claim they help sufferers of insomnia, anxiety and depression; again, they throw the placebo effect in for free. If I suffered from insomnia before a race (or indeed at any time) then I wouldn't care how it worked, I'd just be happy that it did.

Augmentation

For the next few generations or so, sporting performance will not improve by very much. Data shows that in athletics, almost every event is reaching an equilibrium. The growth in performance following the Second World War was driven by a number of things: improvements in nutrition; better coaching and sports science support; an increase in the athlete population; improvements in technology.

I used to worry that the technical innovations I was working on didn't make any difference, but luckily historical data shows they do. Hollow javelins improved performance by four to five per cent between 1953 and 1956. The introduction of composite poles for the pole vault improved performance by eight per cent between 1956 and 1972, and fully automated timing reduced performance by about five per cent when it was introduced in 1975.[147]

This is where I have to introduce the D-word: doping. It's well known that doping has been prevalent in sport in the past, but by how much? Our data shows that performances in running events diminished after the introduction of random controlled drugs testing in 1989. In the women's 200 metres, it dropped by three per cent. The drop was even higher in women's field events: in the shot put, performance dropped by a massive 11 per cent. Drugs testing was strengthened further when the World Anti-Doping Agency was formed at the end of 1999 and performances dropped accordingly.[149] After the post-drug testing falls, the performance in most sports stagnated and showed little in the way of improvement. This suggests that drugs testing is having an effect. It also suggests that we are reaching the limits of human performance with the athletes we currently have.

If this is the case, then where might athletes look for their next performance improvement? One suggestion is stem cell therapy. This is used to help athletes come back from what might previously have been career-ending injuries. It's been reported that tennis player Rafael Nadal (back and knee), LA Lakers basketball player Kobe Bryant (knee) and quarterback for the Colts and Broncos Peyton Manning (neck) all used stem cell therapy during their rehabilitation.

Stem cells are the building blocks of the body and can divide to produce more stem cells or be manipulated to make cells of a different type. They are found in an adult's bone marrow, fat tissue and blood. In stem cell therapy, tens of thousands of your own stem cells are harvested and injected directly into the region of an injury. The stem cells divide to create new muscle, cartilage or bone, and help decrease the inflammation in the injury. The number of cells can be increased by allowing them to grow outside the body for a couple of weeks, producing around 2 million cells. Even this huge number of cells is still tiny, however, weighing less than a hundredth of a gram.

The World Anti-Doping Agency (WADA) allows stem cell therapy to be used, but only if it is applied to the injury. But what if it's applied to non-injured areas? What if a huge number of stem cells is injected? What if the genes of the cells are modified?

At the moment, a report from the US National Academies of Sciences, Engineering and Medicine says that human performance enhancement using stem cells is unlikely because of the significant challenges in growing large numbers of cells.[150] This doesn't mean some won't try, or that WADA isn't watching closely.

One of the key facts about performance is that it improves if the athlete population increases since there is a better chance of finding exceptional athletes. This is why bigger countries tend to do better at the Olympics (India is a clear exception). Before the 1980s, only a handful of the top 25 athletes in endurance running came from Africa; now almost all of them do. And with this influx of African runners, times have dropped. As the athlete population for an event increases, it tends to become more specialised. One hundred-metre sprinters are getting taller and heavier, endurance runners smaller and lighter. While genetic mixing of large populations means that outliers with extreme talents and physiques are more likely, what if we could shortcut the natural selection process and change our genes directly?

Scientists in La Jolla, California modified a mouse with a gene to increase the amount of Type I muscle fibres in its body. Type I fibres use oxidative metabolism for energy production, which makes them fatigue-resistant and good for endurance. The modified mouse was able to run 67 per cent longer and 92 per cent further.[151] Richard Hanson and Parvin Hakimi from Case Western Reserve University in Ohio went on to create a 'supermouse' by modifying the PEPCK-C enzyme; this has a key role for metabolism in the liver, kidneys and fat tissues. The supermouse could run over one kilometre per hour for five hours without stopping. It lived longer than a normal mouse, had sex well into old age and ate 60 per cent more food than an ordinary mouse without putting on weight.[152]

This sounds really quite inviting, and there are undoubtedly some who would love to create a similar superhuman. Some think it has already been attempted: German coach Thomas Springstein wanted to improve oxygen levels in the blood using Repoxygen, a gene therapy used to treat people with anaemia.[153] There's no evidence he ever managed it, and he was convicted for giving drugs to minors. Before the Beijing Olympic Games in 2008, a clinic in China was caught on camera trying to sell performance-enhancing gene therapies for $24,000. The doctor claimed the therapy enhanced lung function and increased the number of stem cells in the blood. Whether anyone had taken him up on the offer or whether the therapy actually worked wasn't known.[154]

Some athletes may take the precarious step of genetic doping to improve performance despite the high risk of contracting serious diseases such as leukaemia. But WADA has worldwide research figuring out how to detect this form of doping, and the promise of the supermouse being translated into a superhuman has yet to be realised (as far as we know).

Challenging the equilibrium

For the moment, then, we might be safe from those who want to cheat through gene or stem cell therapy. But we've got ourselves into a tricky spot: the Olympic motto, *citius, altius, fortius*, means 'faster, higher, stronger', with the implication that performances will always improve. As the century progresses, we're not likely to see improvements in sport unless something changes: equilibrium will have been reached and world records will become scarce.

How can we challenge the equilibrium? What are we going to change? How will we satisfy our innate desire for progression? If our modern Olympic Games are to last as long as the ancient Greek ones did, then the CCLXXXVI Olympic Games will be held in 3036. I expect they'll take place in my home city of Sheffield. What might sport look like then? How might technology have changed performance?

Our data shows that sports tend to change rapidly when a new rule is put in place. Take the javelin: in 1984, Uwe Hohn of East Germany threw it 104.8 metres. It was such a big throw that it landed only a couple of metres short of the track on the opposite side of the stadium. This was the last straw for the IAAF, who had

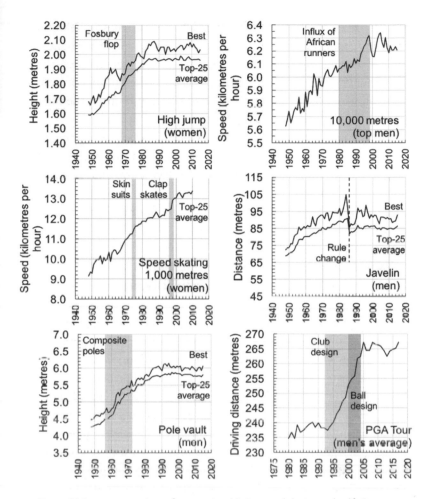

Figure 44. Improvements in performance in athletics, speed skating and golf. *Data courtesy of Leon Foster; golf data adapted from the R&A Distance Report.*[148]

been struggling to adjudicate on the way the javelins landed: they had so much aerodynamic lift that they were landing flat rather than tip first, and it was becoming difficult to tell when the tip touched the ground first.

The IAAF put in a new rule in 1986 to move the centre of mass four centimetres forwards to force the javelin to always land on the point and average distances dropped by a massive ten per cent. Then an interesting thing happened. The javelin throwers and their coaches adapted to the new rules and new javelins, and performance improved, even faster than before the rule change. They tried roughening up the surface of the javelin to alter the lift and drag, just like the dimples on a golf ball. The IAAF brought in another rule to outlaw this too.

Simply changing the rules of a sport is enough to challenge the performance equilibrium. Another way to do this is to create variants of our common sports: three-on-three basketball; mixed triathlon; mixed relays. This isn't fantasy – these will be in the Olympics in Tokyo in 2020. Athletes and coaches will adapt quickly and new world records will be set.

Another sport included in 2020 is sports climbing. This isn't the 'North Face of the Eiger' type of climbing, but action-packed climbing up specially constructed walls. It's still a race, but upwards. Every year in Sheffield we have a festival in the city centre called 'Cliffhanger' where the country's best climbers converge into a packed-out marquee to race up a wooden wall. The wall is 25 metres across and seven metres high, with coloured hand-holds forming different climbing routes so that it looks like a lurid piece of contemporary art. The climbers come out together to stand along the wall with their backs to it. At the appointed time, they turn to see the route they have to climb and up they go. It's a combination of strength, agility and vertical orienteering. Climbers do all the routes and the fastest one overall is the one that wins. It's one of the most tense and exciting sports I've seen for some time, and even my 80-year-old mother enjoyed it.

Two youth sports that have been ripe for entry into the Olympics are surfing and its land-based cousin, skateboarding; again, both have been included in Tokyo 2020. New world records will be set, and I can't wait for the technological discussions about skateboard wheels, climbing shoes and surfboard design. Perhaps UK Sport will ask us to do some research in Hawaii?

Looking further into the future, gaming – or eSports as some call it – is starting to create its own Olympic-style competitions. It will be included in the 2022 Asian Games, and discussions are being held about inclusion in the Paris Olympic Games in 2024. Popular games tend to be the 'shoot-'em-up' strategy style, although there are sports games such as eMLS in soccer and NBA 2K League in basketball. But the current 65-year-old President of the IOC, Thomas Bach, was less than impressed about eSports when asked. While visiting Silicon Valley in 2016, he met an eSports advocate who proudly announced that since the creation of his particular game, 400,000 virtual cars had been destroyed. Bach told *Inside the Games*, "Quite frankly, this did not impress me very much."[155]

The main issue is whether eSports are really sport. Sitting with your hands glued to a games controller is not perhaps how some would define sport. But eSports could easily change this if they took the approach of 'serious gaming', where virtual sports and exercise are integrated. A virtual bike game might be linked to a real static bike so that energy expenditure on the real bike is reflected in the virtual world. Zwift already does this: cyclists all over the world sit on indoor trainers in their garages, spare rooms and home gyms, and are linked together across the internet. They compete in virtual races with live commentary. People are as serious about this as they are about their real races.

How will sport evolve as we fast-forward to the Sheffield Olympic Games of 3036? Since sport reflects people, then we might first ask ourselves what we might be like. Many predictions say that we will continue to grow in height over the next millennium, and that globalisation will cause the population to become more uniform. The number of languages will probably drop from the 7,000 we have today to a few hundred as access to the internet covers the globe. Already, a third of girls born today in the UK will live to be 100 years old: by 3036 the average lifespan is predicted to be around 120. Athletes are likely to be taller, older, look similar and speak the same language.

I spoke to an orthopaedic surgeon, and he said that he'd recently seen a change in the philosophy of the patients he treated. It wasn't good enough to just fix their bones and joints, they now expected to be able to play sport well into their old age. In 2014, 42-year-old Michael Rix had a hip replacement using a new ceramic-coated hip joint made by Sheffield company JRI Orthopaedics. Just three weeks later, he won a triathlon and then went on to represent Great Britain in his age category: he won silver.[156] A time will inevitably come when there will be a discussion similar to the one we had about Oscar Pistorius: when does an implant go from rehabilitation to performance enhancement? Replacement limbs already have robotic joints. Will implants go the same way?

The US National Academies has the brain-computer interface down on its watch list as a future way to enhance human performance. At the moment, computers are poor facsimiles of the human brain: the mighty K-computer from Fujitsu might be four times faster and hold ten times more data than the brain, but it takes almost ten megawatts to power compared to the tiny 20 watts of the human brain. But, the way things are progressing, data and information will be all around us in everything we wear and everything we interact with. The computing power of the K-computer will probably remain remote and we will interact with it wirelessly via small devices, either worn or implanted. If these become the norm in society, sport will have to come up with a set of rules to cope. The International Tennis Federation allows sensors in rackets and equipment, as long as they can't be used for coaching during a game. Will they deal with human-computer augmentation in the same way?

Population sport

As world populations become more similar in the way they behave and speak, popular sports will attract large numbers of people. Take 'parkrun', for example. In 2004, Paul Sinton-Hewitt was injured and decided to design a time trial as a way to stay involved in running and, most importantly, keep in touch with his friends who were runners. He measured out a five-kilometre course in Bushy

Park in London and told his friends he would be their timer if they fancied it. He brought a stopwatch, a table and a pad of paper: thirteen turned up for the first event.

Before long, friends of friends began to take part and word of mouth spread the appeal of this free, volunteer-led time trial. Enthusiasm for the concept was such that a second time trial was set up five miles away on Wimbledon Common and then another one 200 miles away in Leeds. These time trials began to attract others who weren't there to get a best time but were just happy to get around. Anyone from age four upwards was allowed to participate, whether running or walking. Now there are over 1,600 parkruns across the world with over three million people having taken part. Aged 92, Norman Phillips completed his 100th parkrun in 2016.

The appeal of parkrun is simple: it's at the same time, same place, same format every weekend. Some use it to gauge their fitness, some use it to get fit, some use it just to meet friends. Technology is an important part of parkrun, but it remains simple: as you run, walk or totter your way past the finishing line, someone gives you a token and you queue up for a volunteer to scan it, together with a preprinted barcode you bring with you. This has your unique parkrun number which allows you to go online later to see your results and compare yourself to those of your friends and competitors. For the record, my number is A88020.

The queues get quite long at the popular runs – Bushy Park regularly has more than 1,200 runners, my local one in Sheffield has over 700. Technology could time everyone automatically as they crossed the line, of course, but parkrun has realised that this would be its downfall. Queuing is part of what makes parkrun special: people talk to each other, compare their times, ask how each other are and become lifelong friends. There are parkrun rivalries, romances, weddings and births.

parkrun will soon have a million runners and walkers every weekend, the world's first population-level sporting activity. By the year 3036, parkrun may have the whole world turning up each weekend. Equally, it may have run its course and been replaced

Figure 45. The search for nostalgia: modern penny-farthings at the Great Manchester Cycle Ride in 2014. Difficult to get into the bike park. © *Steve Haake*.

by something else. The message, however, is that sport is not just about competition, it's now about social connections. parkrun does this very well, as do Zwift, Strava, MapMyRun and a whole host of other online communities.

In 3036, the world will be data rich with information swirling all around us. I suspect there will be nostalgia for simpler times and, as populations begin to look very similar, there will be a desire to be different. I saw this happen recently at a bike race. As I waited for my son to cross the line, four penny-farthings raced in. These weren't old penny-farthings, they were brand new, complete with cleated pedals and bike computers. In a world where technology is there to make things easier, here were bikes to make things harder. This might be another clue to where sport will go in the next 1,000 years: old challenges made new.

Back to square one[157]

If someone were to write this book again in 3036, I wonder which technology they would think changed the history of sport the most. Materials are often the catalyst for our designers to come up with something new: the jumping weights of the ancient Greeks showed that they understood well the principles of design for performance using lead, bronze and stone. Howard Head used metal, polymers and then carbon fibre to design his oversized tennis racket, as did Van Phillips when he created his carbon fibre prosthetic. Speedo's swimsuits only existed because of the development of Lycra and polyurethane textiles, and even the St Moritz bobsleigh track was the melding of materials and design: ice for the track, steel for the runners. I'm sure that graphene will be equally influential.

But, in the rapid rise of world sport between the mid-19th century and the present day, there is one technology that stands out for me. It single-handedly created most of world sport and without it, we wouldn't have the Super Bowl, the FIFA World Cup, the Six Nations, the Sam Maguire Cup and Wimbledon. This is what I think will have had the biggest effect on world sport, even in 3036: vulcanised rubber.

So, thank you to Charles Goodyear for your passion and your obsession. But let's not ignore the others who had their own technological influence on sport: Major Walter Clopton Wingfield, Howard Head, Eadweard Muybridge, Étienne-Jules Marey, Harold Edgerton, Robert Paterson, Coburn Haskell, Peter Guthrie Tait, William Taylor, the Outdoor Amusements Committee of the Kulm Hotel, Annette Kellerman, Wallace Carothers, Alexander MacRae, Karl von Drais, James Starley, William Hillman, Chris Boardman, Graeme Obree, Mike Burrows, Franz Krienbühl, Gerrit Jan van Ingen Schenau, Ludwig Guttmann, Bob Hall, Rainer Küschall, Van Phillips, Yoshiro Hatano and Seppo Säynäjäkangas.

Every distant future I've ever seen in film has flying cars. The ancient Olympics started with a simple race and ended as a complex tournament that included chariot racing. Does this mean we'll have flying cars in the 3036 Olympics? It might seem far-fetched, but at a meeting in Texas recently, someone proudly told me of two-person drones being developed to commute short distances into Dallas-Fort Worth airport. I guarantee that as soon as this happens, there'll be a race. Sport might evolve with the technology and moral values of the day, but our desire for competition will always remain the same. I have no doubt that 1,200 years from now, we'll still be scratching two metaphorical lines in the sand and racing between them, just as our ancient Greek ancestors did 2,700 years ago.

Steve Haake
AD 2018

PLAYLIST

THANK YOU to BBC 6Music for keeping me going while I was writing; the BBC iPlayer is a wonderful invention. Thanks to all the presenters, especially to Steve Lamacq, Mark Riley, Gideon Coe, Mary-Ann Hobbs, Cerys Matthews and Tom Ravenscroft. Not everything in this list comes from you, but much of it does.

I can't claim that I was listening to these tracks as I actually wrote – often that required complete silence. And anyway, good music deserves your full attention (apologies to all artists when I don't give it). But writing needs a lot of thinking time and bucket loads of inspiration: these gave me both. If you want a sports compilation, try *It's the taking part that counts: a global pop sportsday* from WIΛIWYΛ. If you're interested, here is my playlist, chapter by chapter.

Chapter 1. Lucia di Lammermoor; Donizetti, Philharmonia Orchestra and Chorus. This record seemed to have the grandeur that a chapter about the ancient Greeks required and, contrary to what I just said, did actually allow me to write while I listened.

Chapter 2. The Bees; Every Step's a Yes. Thinking about ancient Greece and the hot sun led me to this album by the Bees; it always makes me think of summer.

Chapter 3. Destroyer; Poison Season. If you imagined heavy metal because of the band name, think again. Dan Bejar's album is bright and busy, even if the words are sometimes slightly disturbing. It seemed to fit a chapter with its own share of contradictions.

Chapter 4. King Creosote; Astronaut meets Appleman. Every year, about the time I was writing this chapter, we go up to Kirkcudbright in Scotland to visit friends and do the half marathon. That's probably why I was listening to Scotsman Kenny Anderson's jangly guitars.

Chapter 5. 10CC; Live and Let Live. A blast from the past which matched my reminiscences in this chapter from thirty years ago. Made me go and see them live twice during the year – they're still as good as ever.

Chapter 6. Orange Juice; Rip it up. Again, thinking about writing my PhD in the 1980s led me back to this, a wonderfully bouncy album from the magnificent Edwyn Collins and his band.

Chapter 7. Okkervil River; Away. The winter sport of this chapter reminds me of being in the cold when I was filming with Kensington TV; it makes me want to blow on my hands to warm them up. This album matched that feeling: melancholy and atmospheric.

Chapter 8. Darren Haymen; Lido. An album of instrumentals about London's open air swimming pools. Perfect for a chapter on swimming, the vinyl is bright blue and the sound is pure British summer.

Chapter 9. The Shins; Heartworms. My kids introduced me to the Shins, and this eclectic album seemed to match the complexities of the chapter.

Chapter 10. Whitney; Light Upon the Lake. Perhaps it's the cover and the title, but it reminds me of being in Canada with Kensington TV, filming winter sports in the cold. I love the twiddly guitar bits matched by the sharp brass.

Chapter 11. Public Service Broadcasting; The Race for Space. What better way to engage with 20th century history than with PSB? *Come on guys, let's do sport!* Their concerts get better and better. (Apologies for the idiot in the crowd at the Sheffield gig.)

Chapter 12. Todd Terje; It's Album Time. A stunning album of electronic pop to go with the technology of the age that this chapter represents.

Chapter 13. Hannah Peel; Mary Casio: Journey to Cassiopeia. This seemed a fitting album for a peer into the future and is one of the most beautiful things I own with its shiny silver gate-fold sleeve. Her concert at the Queen Elizabeth Hall with a full brass band is probably the most emotional one I've ever been to (yes, that's a euphemism for 'I cried').

BIBLIOGRAPHY

THERE ARE many, many books about sport, its heroes, its culture and its history. Finding out about how sports technology developed required me to line up all the sports books I could find and then dive into each to see what pockets of knowledge were inside. There were, of course, academic journals and conferences, two of which I set up myself during the 1990s to collect the disparate articles on sports engineering that were already going to publications where they didn't really belong.

This bibliography shows the major texts that I used during my research for this book; there are many more I've not listed. Some gems are only available in second-hand bookstores or at the click of a mouse. I had to stop buying books and go to the library instead when I could no longer see out of my study window.

Academic journals and books on the physics and engineering of sport

Daish, C.B., *The Physics of Ball Games* (London, Hodder & Stoughton, 1981). Fittingly, the first in this bibliography is the one that most people interested in the physics of sport first turn to. Sadly out of print, it was the first to try to bridge the gap between a textbook and a popular science book. If you find a copy, you're very lucky indeed.

Frohlich, C. (ed.), *Physics of Sports: Selected Reprints* (American Association of Physics Teachers, 1986). This was one of the first books I bought when I began teaching. It's a neat collection of difficult-to-find papers if you're not an academic. A perfect volume on the physics of sport.

Haake, S.J. (ed.), *The Engineering of Sport*, Proceedings of the 1st International Conference on the Engineering of Sport, Sheffield, UK, 2–4 July 1996 (Rotterdam, Balkema, 1996). This contains 50 or so papers from the first international conference dedicated to sports engineering. There were further conferences in Sheffield (Haake, 1998), Sydney (Subic and Haake, 2000), Kyoto (Ujihashi and Haake, 2002), Davis (Hubbard, Pallis and Mehta, 2004) and Munich (Moritz and Haake, 2006). The latest conference was The Engineering of Sport 11 held in March 2018 in Queensland, Australia.

Haake, S.J. (ed.), *Sports Engineering, vol. 1*, 1998 to today. This was the first academic journal on Sports Engineering and contains hundreds of peer-reviewed papers on science, technology and sport. I edited the first six volumes.

Ancient Greek sport

Gardiner, E.N., *Athletics of the Ancient World* (Oxford, Oxford University Press, 1971). First published in 1930, this is one of the earliest summative accounts of athletics in the ancient world, with theories on *halteres* and a few anecdotes to spice up the old-fashioned writing.

Miller, S.G., *Ancient Greek Athletics* (New Haven and London, Yale University Press, 2004). With 291 black and white photographs, this is probably the finest book you can buy on the athletics of the ancient Greeks. A book to return to again and again.

Perrottet, T., *The Naked Olympics: The True Story of the Ancient Games* (London, Random House, 2004). A humorous and irreverent look at what it was probably like to experience the ancient Olympics first hand. I gave this to a serious Olympic scholar once; I don't think she was impressed because she never mentioned it again. It's a great read.

Swaddling, J., *The Ancient Olympic Games* (London, The British Museum Press, 2004). A book I picked up in the shop of The British Museum; it's a quick introduction to some of the themes of the ancient Olympic Games dealt with more deeply by Gardiner and Miller.

Sweet, W.E., *Sport and Recreation in Ancient Greece* (Oxford, Oxford University Press, 1987). A sourcebook with translations. A good place to find the original quotes that all the others refer to.

Valavanis, P., *Hysplex: The Starting Mechanism in Ancient Stadia* (Berkeley, The University of California Press, 1999). This is not for the faint-hearted; it's an in-depth study of the starting gates of ancient Greece and from which you could almost make your own back-garden *hysplex*.

Biographies

Gibson, E. and Firth, B., *The Original Million Dollar Mermaid: The Annette Kellerman Story* (Crows Nest, Australia, Allen and Unwin, 2005).

Annette Kellerman's story is not quite captured by the film of a similar name, but told here by fans, the Australian heroine is just amazing.

Lewis, M., *Moneyball: The Art of Winning an Unfair Game* (New York, W.W. Norton & Co. Inc., 2004). The book that set the sports world off in search of the analytics Holy Grail.

Pistorius, O., *Blade Runner* (London, Virgin Books, 2012). Ghostwritten by Gianni Merlo, this is a strange look into the mind of Pistorius. It is interesting for Pistorius' self-aggrandisement and insular view of the world.

Slack, C., *Noble Obsession: Charles Goodyear, Thomas Hancock and the Race to Unlock the Greatest Industrial Secret of the Nineteenth Century* (London, Texere Publishing, 2002). I'm not sure whether the title refers to the subject, Charles Goodyear, or the author himself but this gives an incredibly detailed account of the discovery of the vulcanisation of rubber.

Bobsleigh

Triet, M., *100 Jahre Bobsport* (Basel, Schweizerisches Sportmuseum, 1990). Consisting of a 104-page book and a 120-page catalogue, these are two of my favourite books: the smaller one is essentially a book on 'A century of bobsleighing' while the longer one is an exhibition catalogue with the same name accompanying an exhibition that was held in 1990.

Cycling

Herlihy, D.V., *Bicycle* (New Haven and London, Yale University Press, 2004). A lovely book with history, culture, pictures and a love for all things cycling.

Hutchinson, M., *The Hour: Sporting Immortality the Hard Way* (London, Yellow Jersey Press, 2007). A great book that tells you all you need to know about trying to cycle the hour; so enjoyable it stopped me writing until I'd finished it.

Ritchie, A., *King of the Road: An Illustrated History of Cycling* (Berkeley, Ten Speed Press, 1975). An in-depth look at the history of the bicycle, worth getting just for the pictures not in other books (this Andrew Ritchie is not to be confused with the other Andrew Ritchie who created the Brompton folding bicycle).

Whitt, F.R., and Wilson, D.G., *Bicycling Science* (Cambridge, Massachusetts, MIT Press, 1982). The seminal scientific text on cycling; the first book any serious scientist working in cycling picks up.

Football

Anderson, C. and Sally, D., *The Numbers Game: Why Everything You Know About Football is Wrong* (London, Viking, Penguin Group, 2013). Football's answer to *Moneyball*, except the stars of this book are the numbers and the authors' stories rather than the manager of a baseball team. One of my best holiday reads.

Davies, H., *Boots, Balls and Haircuts: An Illustrated History of Football From Then 'til Now* (London, Cassell Illustrated, 2003). A self-confessed Tottenham Hotspur fan, Davies gives an entertaining blast through the history of football. A great read and insight into the technical culture of soccer through the ages.

Inglis, S., *A Load of Old Balls* (Bristol, English Heritage Publishing, 2005). I just love this book: small, easy to read and says so much in such a small space.

Golf

Browning, R., *A History of Golf: The Royal and Ancient Game* (London, A&C Black (Publishers) Ltd., 1990). Originally published in 1955, it's the brain-dump from the editor of *Golfing* magazine. A view of golf from 60 years ago.

Cochran, A. and Stobbs, J., *The Search for the Perfect Swing* (The Golf Society of Great Britain, 1968). Still in print (in an internet sort of way), this is probably still the best book if you want to understand the science of golf. Its only downfall is that the technology stops in 1968.

Hotchkiss, J.F., *500 Years of Golf Balls: History and Collector's Guide* (Iowa, Antique Trader Books, 1997). For the golf nerds amongst us, this tells you everything you need (and don't need) to know about golf balls. An intriguing look inside the golf industry.

Motion capture, photography and biomechanics

Braun, M., *Eadweard Muybridge* (London, Reaktion Books, 2010). A short study of Muybridge, not as lovingly crafted as her work on Marey (see below) but useful for a comparison of the styles of the two frontiersmen.

Braun, M., *Picturing Time: The Work of Étienne-Jules Marey 1830–1904* (Chicago, The University of Chicago Press, 1992). This introduced me to the beauty and elegance of Marey's work and its comparison to the rather slipshod style of Muybridge. A fantastic history of technology and the life of a great scientist.

Brookman, P., *Eadweard Muybridge* (London, Tate Publishing, 2010). A huge book with contributions from four different authors, each giving their thoughts on Muybridge the man, the artist and the scientist. As you would expect from a book to go with a Tate exhibition, the 194 photographs are sublime.

Jussim, E., *Stopping Time: The Photographs of Harold Edgerton* (New York, Harry N. Abrams Inc., 2000). A collection of Edgerton's photographs printed in high quality. His breadth of work was astonishing.

Marey, E-J., *Animal Mechanism: A Treatise on Terrestrial and Aerial Locomotion* (New York, D. Appleton and Company, 1879). Marey's original book that inspired Muybridge and others to use photography for serious science. The graphs and pictures were the first of their kind. Still available on the internet.

Performance

Epstein, D., *The Sports Gene: Talent, Practice and the Truth About Success* (London, Yellow Jersey Press, 2013). A giant of a book with data, stories and references; hard to argue against.

Slot, O., *The Talent Lab: The Secrets of Creating and Sustaining Success* (London, Ebury Press, 2017). An insider's view of Britain's Olympic success by the performance directors of UK Sport, Simon Timson and Chelsea Warr.

Sports history

Guttmann, A., *Sports: The First Five Millennia* (Amherst, Massachusetts, The University of Massachusetts Press, 2004). A comprehensive look at the history of sport since the Egyptians. It's quite academic in style and gives a useful definition of sport.

Holt, R., *Sport and the British: A Modern History* (Oxford, Oxford

University Press, 1989). A deep depiction of the history of sport in Britain and its effect on the world.

Huggins, M., *The Victorians and Sport* (London, Hambledon and London, 2004). A detailed account of the growth of sport against the background of Victorian Britain. Sets world sport in context.

Tranter, N., *Sport, Economy and Society in Britain 1750–1914* (Cambridge, Cambridge University Press, 1998). Really dense stuff but full of interesting facts that just ask to be followed up.

Tennis

Brody, H., *Tennis Science for Tennis Players* (Philadelphia, University of Pennsylvania Press, 1987). This is the original book on tennis science by its foremost expert. He was the Feynman of tennis and players; scientists and those in charge of tennis loved him.

Collins, B., *Total Tennis: The Ultimate Tennis Encyclopedia* (Toronto, Sport Media Publishing Inc., 2003). A huge book, literally. Detail on every event in tennis up to 2002; the only way to find out what happened way back when.

Gillmeister, H., *Tennis: A Cultural History* (London, Leicester University Press, 1997). Gillmeister provides a rich understanding of the origins and development of tennis. Not a book to be argued with.

Maxton, P., *From Palm to Power: The Evolution of the Racket* (London, Wimbledon Lawn Tennis Museum, 2008). A neat little introduction to the development and manufacture of tennis rackets.

NOTES

Chapter 1: Starting from scratch

[1] The documentary was called *The Equalizer* and was produced by Kensington Communications (Canada) and Berlin Producers (Germany). It followed five Olympic athletes as they tried to beat Olympic champions from the past using old sports technology. See theequalizertv.com.

[2] Stefanyshyn, D. and Fusco, C. (2004), 'Increased shoe bending stiffness increases sprint performance', *Sports Biomechanics*, 3(1), pp. 55–66. The researchers analysed 34 sprinters with four different stiffnesses of sole plate. The performance of 29 of them improved with a stiffer shoe, while three didn't improve and two got worse. The best performance improvement was over three per cent, but the optimal stiffness didn't correlate to any measures such as height, weight or shoe size.

[3] Foster, L. (2009), 'The effect of technology on elite performance', PhD thesis, Sheffield Hallam University.

[4] Haake, S.J., Foster, L.I. and James, D.M. (2014), 'An improvement index to quantify the evolution of performance in running', *Journal of Sports Sciences* 32:7, pp. 610–622. The performance improvement

index is a way of comparing different sports. It uses the concept of energy expended during an event; for example, that in throwing or jumping it is proportional to the distance or height, in running it is proportional to the time squared, and in the hour record it is proportional to the distance cubed. See further references below.

[5] bit.ly/1PKkfsw. This is an amazing 1964 documentary featuring Jesse Owens returning to Berlin after the war. It is a dramatic social and sporting commentary of the times.

[6] It is difficult to judge both the start and last frames so there is an error in my calculation of ±1 frame at each end, adding up to ±2 frames. Since one frame is 0.033 seconds, the error is equal to about 0.07 seconds. The best I can quote my measurement to is 10.63 ± 0.07 seconds. My measurement error is still smaller that the judge's reaction time.

[7] McMahon, T.A. and Greene, P.R. (1979), 'The influence of track compliance on running', Journal of Biomechanics 12, pp. 893–904.

[8] Farley, C.T. and Gonzalez, O. (1996), 'Leg stiffness and stride frequency in human running', Journal of Biomechanics 29, pp. 181–186. Also Mann, R., Kotmel, J., Herman, J., Johnson, B. and Schultz, C. (1984), 'Kinematic trends in elite sprinters', 2nd International Symposium on the Biomechanics of Sport, Colorado Springs.

[9] Ancient scholars such as Aristotle agreed on a list of victors in the stade sprint and counted back in units of four years. In the modern Gregorian calendar, the first race took place in 776 BC. There may have been events back beyond this, but they weren't recorded so this date is generally taken as the first ancient Olympic Games.

[10] Dr Jason König, from the University of St Andrews, has a great blog on differences and similarities of the ancient and the modern Olympics, including a lovely piece on the concept of sweat collectors. http://ancientandmodernolympics.wordpress.com/.

[11] Miller, S.G., *Ancient Greek Athletics* (New Haven and London, Yale University Press, 2004), p. 42.

[12] The *hysplex* can be seen in action at nemeangames.org.

[13] Miller, *Ancient Greek Athletics*, p.134. A skilled Greek artisan was paid around one drachma a day, which is approximately equivalent to $100 per day in today's money.

[14] The Pose Method ® of running has many advocates across the world, not least Nicholas Romanov who wrote a book on the subject and sells educational seminars about the technique.

[15] De Wit, B., De Clercq, D. and Aerts, P. (2000), 'Biomechanical analysis of the stance phase during barefoot and shod running', *Journal of Biomechanics* 33, pp. 269–278.

[16] BBC News 8 May 2014. Vibram in $3.75m settlement over false health claims. *www.bbc.com/news/business-27335251*.

[17] The race is at http://bit.ly/2tbmDJE. I'm fourth from the right in lane Δ.

[18] The speed of sound is 343 metres per second at 20°C. Sound would take three hundredths of a second to travel from the gun to the runner in the outside lane, ten metres away from the starter standing inside the track. In contrast, the runner in the inside lane would hear it after less than a hundredth of a second.

[19] Van Ingen Schenau, G.J., De Koning, J.J. and De Groot, G. (1994), 'Optimisation of sprint performance in running, cycling and speed skating', *Sports Medicine* 17(4), pp. 259–275. The conclusion was actually for speed skate sprinters, but the principle is similar for runners.

Chapter 2: The shape of things to come

[20] Harris, H.A. (1960), 'An Olympic Epigram: The athletic feats of Phayllos', *Greece & Rome* vol. 7(1), pp. 3–8. Harris' piece reads like it emerged from a heated discussion in the senior common room of St John's College, Cambridge, possibly over a fine port.

[21] *The Oxford Encyclopedia of Ancient Greece and Rome* (2010), p. 382, gives the distances between the starting lines of a number of stadia: Olympia 192.29 m; Epidaurus 181.18 m; Isthmia 181.20 m; Nemea 178.02 m; Delphi 177.41 m; Halieis 166.50 m.

[22] I've not managed yet to find a paper that proves the link between improved performance and increased ground reaction forces due to the extended swinging of the arms. However, this is a common perception.

[23] Brody, H. (1985), 'The moment of inertia of a tennis racket', *The Physics Teacher*, pp. 213–216. Consider a tennis racket with a mass M split up into regular sections of mass δm at a distance r from the handle end. The three mass moments of inertia, I_j, are defined by $I_j = \sum_{i=0}^{n} \delta m_i r_i^j$, where n is the number of mass elements and $j=0, 1, 2$. When $j=0$, then $I_0 = \sum_{i=0}^{n} \delta m_i = M$, i.e. the mass of the racket. When $j=1$, $I_1 = \sum_{i=0}^{n} \delta m_i r_i = MR$ where R is the distance from the handle to the centre of mass of the racket; this is the downward moment felt by the hand when the racket is held horizontally by the handle. When $j=2$, $I_2 = \sum_{i=0}^{n} \delta m_i r_i^2 = MR^2$; this is the moment of inertia of the racket or its 'swingweight'. It is felt as the resistance to rotational motion when swinging it about the handle. A more common way of writing this is $I = \frac{1}{3}ML^2$ where L is the length of the beam.

[24] Kron, G. (2005), 'Anthropometry, Physical Anthropology, and the Reconstruction of Ancient Health, Nutrition and Living Standards', *Historia: Zeitschrift für Alte Geschichte* 54, H.1, pp. 68–83.

[25] Plagenhoef, S., Evans, F.G. and Abdelnour, T. (1983), 'Anatomical data for analysing human motion', *Research Quarterly for Exercise and Sport* 54, pp. 169–178. Stanley Plagenhoef, from the University of Massachusetts-Amherst measured athletes in the 1980s to find the mass and length of the limbs for his own biomechanical models. He found that an athlete's arm is always about five per cent of his weight and about 40 per cent of his height. This would have made Phayllos' arms about 56 centimetres long and weigh three and a half kilograms. His hand would have weighed around half a kilogram. This allowed me to estimate the moment of inertia of his arm when swung about his shoulder. A concentrated weight of 1.2 kilograms would have had the same moment of inertia. See here for the anthropometric data: http://www.exrx.net/Kinesiology/Segments.html.

[26] Minetti, A.E. and Ardigó, L.P. (2002), 'Halteres used in ancient Olympic long jump', *Nature* 420 (69), pp. 141–142. They showed that the weights slowed down the swing speed of the arms which slowed down the loading of the muscles in the lower body. Muscles are more efficient at lower loading rates, and so the ground reaction force increased.

[27] How far could Usain Bolt jump? http://engineeringsport.co.uk/2010/03/03/how-far-could-usain-bolt-jump/.

[28] In Homer's *Iliad*, Achilles introduces the discus to his soldiers: "Now men come forward – compete to win this prize! An ingot big enough to keep the winner in iron for five wheeling years."

[29] Murray, S.R., Sands, W.A., O'Roark, D.A. (2011), 'The ancient

Greek dory: how effective is the attached ankyle at increasing the distance of the throw?', *Palamedes* 6, pp. 137–151. Murray and colleagues seem to have used the wrong camera frame rate in their calculation of velocities, quoting launch speeds of about four to five metres per second. This is very small and equivalent to the speed a ball would reach if dropped from shoulder height; a javelin at this speed would only have reached about five metres away, not much better than my effort with my mum's broad bean poles. I'm reasonably confident that their distances are correct, however, as these were measured independently.

Chapter 3: Starting from scratch

[30] Miller, M., 'The Maya ballgame: rebirth in the court of life and death' in *The Sport of Life and Death* ed. by M.E. Whittington (London, Thames and Hudson, 2001), pp. 79–87. The Popul vuh is the Mayan creation myth and tells the story of the creation of man, based around the Mesoamerican ballgame. The heroes are the twins Hunahpu and Xbalanque who play a ball game to avenge the deaths of their father and uncle by the Lords of the Underworld. The heroes win, ascend into the sky and become the Sun and Venus, or possibly the moon.

[31] Hosler, D., Burkett, S.L. and Tarkanian, M.J. (1999), 'Prehistoric Polymers: Processing in Ancient Mesoamerica', *Science* 284, pp. 1988–1991.

[32] Cross, R., (2000), 'The coefficient of restitution for collisions of happy balls, unhappy balls and tennis balls', *American Journal of Physics* 68(11), pp. 1025–1031.

[33] Leyenaar, T.J.J. and Parsons, L., *Ulama: the ballgame of the Mayas and the Aztecs* (Leiden, Spruyt, van Mantgem and DeDoes bv, 1988).

[34] Rühl, J. (2001), 'Regulations for the Joust in Fifteenth-Century Europe: Francesco Sforza Visconti (1465) and John Tiptoft (1466)', *The International Journal of the History of Sport* 18:2, pp. 193–208.

[35] Gillmeister, H., *Tennis: A Cultural History* (Leicester, Leicester University Press, 1997), p. 120.

[36] Official Catalogue of the Great Exhibition of the Works of Industry of All Nations 1851. Spicer Brothers, London. Available https://archive.org/details/officialcatalog06unkngoog.

[37] Slack, C., *Noble Obsession: Charles Goodyear, Thomas Hancock and the Race to Unlock the Greatest Industrial Secret of the Nineteenth Century* (London, TEXERE Publishing, 2002).

[38] Dahms, S.E., Piechota, H.J., Dahiya, R., Lue, T. and Tanagho, E.A. (1998), 'Composition and biomechanical properties of the bladder acellular matrix graft: comparative analysis in rat, pig and human'. *British Journal of Urology* 82, pp. 411–419.

[39] http://richardlindon.co.uk/.

[40] As Rugby Football became shortened to 'rugger', Association Football became 'soccer'.

[41] *Encyclopedia of World Sport: From Ancient times to the Present* (1996), Ed. David Lewinson and Karen Christensen, ABC-Clio, Santa Barbara, CA, p.1142. Apologies to volleyball players everywhere. The original game was created for middle-aged, rather stout and unfit businessmen who weren't fit enough for basketball. I've played the game and it's anything but 'easy'.

[42] The Ideal Gas Law can be used to approximate the behaviour of gases. The formula is given by $PV=nRT$ where P is the pressure,

V the volume, T the temperature in Kelvin, R the universal gas constant (8.314 J mol[-1] K[-1]) and n the amount of the gas in moles. For a fixed volume, such as a pumped-up football, the ratios of the absolute pressures and temperatures are approximately equivalent, i.e. $P_1/P_2=T_1/T_2$. Thus, as the temperature drops, so does the pressure. *Note:* if you try the calculation, don't forget to put the temperature in Kelvin (i.e. room temperature is about 294° Kelvin) and the absolute pressure (i.e. the initial ball pressure was 14.7+12.5=27.2 psi).

43 Theodore V. Wells Jr, Brad S. Karp, Lorin L. Reisner (2015) Investigative Report concerning Footballs used during the AFC Championship Game on 18 January 2015. https://www.documentcloud.org/documents/2073728-ted-wells-report-deflategate.html.

Chapter 4: A game of invention

44 Goodwill, S.R. and Haake, S.J., 'Why were spaghetti string rackets banned in the game of tennis?' in Ujihashi, S. and Haake, S.J. (eds.) *The Engineering of Sport 4* (Oxford, Blackwell Science, 2002) pp. 231–237.

45 Shakespeare, W., *Much Ado about Nothing*, Act III, Scene II.

46 Gillmeister, *Tennis: A Cultural History*, p. 104. The guts of cats were never used for the strings of a tennis racket. According to Gillmeister, the name came from the Dutch name for a tennis-like racket game called *kaetsen*. When exported to Germany and Britain, people mistook the word for 'cats' so that when a racket was restrung, the name of the material was kaetse gut, or 'cat-gut'. Natural strings come from the intestines of sheep.

47 Gillmeister, p. 352.

⁴⁸ Trengrove, A., *The Story of the Davis Cup* (London, Stanley Paul, 1985), p. 25.

⁴⁹ Maxton, P., *From Palm to Power: The Evolution of the Racket* (Wimbledon Lawn Tennis Museum, 2008), p. 37. Frank Donisthorpe went on to work for Dunlop in the 1950s where he developed oversized rackets.

⁵⁰ You can find the centre of percussion in the following way: grip a racket vertically in your hand and use a ball to hit it at the highest point of the face. At the top, the head moves backwards so that the handle moves forwards, jarring your hand. Move slowly downwards towards the handle and continue to hit the racket. As you do this, the jarring at the grip vanishes: this is the centre of percussion. I was once told that if a door is jammed, the best place to shoulder it is the centre of percussion about a third of the way from the open edge. This vertical line is the centre of percussion of the door where there is no impulse at the hinges: hitting it here saves you having to fix the hinges later and is also the best place to put the doorstop.

⁵¹ Simpson, B. *Winners in Action: The Complete Story of the Dunlop Slazenger Sports Companies* (Fakenham, J.J.G. Publishing, 2005), p. 190.

⁵² Simpson, p. 189.

⁵³ Haake, S.J., Allen, T., Choppin, S. & Goodwill, S.R. (2007). 'The evolution of the tennis racket and its effect on serve speed', *Tennis Science & Technology*, 3, pp. 257–271, (Ed. S. Miller & J. Capel-Davis).

The astute reader might wonder why the percentage reduction in reaction time isn't the same as that of the reduction in ball speed. The reason is due to the latency in the time for a player to perceive the ball and then start to react. The first serve from a wooden racket took 0.571 seconds to reach the baseline;

the equivalent with a carbon fibre racket took 0.546 seconds, a decrease of 0.096 seconds. According to Vic Braden, a respected tennis coach, a player doesn't perceive the ball until about 0.25 seconds into the flight so that the increase in reaction time available is 0.096/(0.571-0.25)=7.8%.

Chapter 5: Seeing is believing

54 Cochran, A. and Stobbs, J. (1968), 'The Search for the Perfect Swing', The Golf Society of Great Britain.

55 Muybridge was actually born as Edward James Muggeridge. He changed his name at least three times throughout his life, ending with Eadweard James Muybridge; his initials, however, remained the same.

56 Braun, M., *Eadweard Muybridge* (London, Reaktion Books, 2010), p. 14. The original quote comes from the memoirs of his cousin, Maybanke Susannah Anderson.

57 *The Greek Slave* was sculpted by Hiram Powers in 1843 and 1844. The life-sized statue was of a naked young slave girl, chained and put on sale by her captors. The public was initially scandalised but it became a symbol of Christian purity in the face of evil and a symbol for abolitionists of slavery. There is a 3D image of the statue courtesy of The Smithsonian Institution at 3d.si.edu.

58 Marey, E-J., *Animal Mechanism: A Treatise on Terrestrial and Aerial Locomotion* (New York, D. Appleton and Company, 1879).

59 https://www.theguardian.com/sport/2007/jul/11/tennis.wimbledon.

60 Tennis hasn't yet taken the plunge of showing the close-up of

the ball during a replay. There is a natural nervousness about what would happen if the crowd didn't agree with the decision the images represented.

Chapter 6: Taking the rough with the smooth

61 http://golftips.golfweek.com/history-callaway-golf-balls-1456. html.

62 R&A Rules Ltd. and USGA (2008). Initial Velocity Test Procedure. Revision 10-08, www.randa.org/RulesEquipment/ Equipment/Equipment-Submissions/Test-Protocols

63 Reprinted in Hotchkiss, J.F., *500 Years of Golf Balls: History and Collector's Guide* (Iowa, Antique Trader Books, 1997).

64 William Taylor (1905), An improvement in golf balls. British patent 18,668, accepted 26 April 1906. From worldwide.espacenet.com.

65 This part of the golf story is revealed thanks to a little-known organisation called the Narborough and Littlethorpe Heritage Society and Christopher Jones in particular who diligently researched William Taylor. www.narboroughandlittlethorpe. co.uk/articles/.

66 www.franklygolf.com/golf-ball-testing.aspx. Frank Thomas was the technical director of the USGA until 2000. His blog outlines here the transition from the Iron Byron to the Indoor Testing Range which he instigated.

67 Haake, S.J., Goodwill, S.R. and Carré, M.J. (2004), 'A new measure of roughness for defining the aerodynamic performance of sports balls'. Proc. IMechE vol. 221 Part C:J, *Mechanical Engineering Science*, pp. 789–806.

68 R&A and USGA (2017) 2017 Distance report. www.randa.org/News/2018/03/Distance-Report.

Chapter 7: Sliders

69 Produced by Kensington TV in collaboration with Berlin Producers and PreTV, the winter sports version of *The Equalizer* was called *Champions vs Legends*. http://kensingtontv.com/film-detail/champions-vs-legends/.

70 Triet, M., *100 Jahre Bobsport* (Basel, Schweizerisches Sportmuseum, 1990).

71 The mass of the Earth attracts bodies at its surface so that they experience an acceleration of 9.81 metres per second at the equator, or one gravity. An acceleration of four gravities is four times this, the practical outcome of this is that you feel four times heavier.

72 International Bobsled Rules (2015) International Bobsleigh and Skeleton Federation, June 2015, p. 27. The limits for the two-women bobsleigh are 165 kilograms empty and 325 kilograms loaded; for the men it is 170 kilograms and 390 kilograms. For the four man bobsleigh the limits are 210 kilograms empty and 630 kilograms loaded. There is no four woman event.

73 There is debate whether Galileo actually did the experiment and most think it was probably just a thought experiment. Astronaut David Scott repeated the experiment on the moon using a hammer and a feather to show that in the absence of air, two objects do indeed fall at the same rate. https://bit.ly/2sajh8H.

74 http://www.nytimes.com/1984/02/09/sports/bobsled-rivalry-intensifies.html.

75 Huffman, R.K. and Hubbard, M. (1996), 'A motion based virtual reality training simulator for bobsled drivers' in Steve Haake (Ed.) *The Engineering of Sport*. Proceedings of the 1st International Conference on the Engineering of Sport, Sheffield, UK, 2–4 July 1996. Balkema, Rotterdam, pp.195–203.

76 http://www.bobclub-stmoritz.ch/?rub=12 The website is in German.

77 http://www.slideshare.net/NorbertGruen/20140708speedonice A good example of the equations of motion are given by Dr Norbert Grün and his work with the Institute for Research and Development of Sports Equipment in Berlin (FES) on the development on bobsleighs. The Germans topped the medal table in bobsleigh in Sochi 2018 with three gold medals and one silver medal.

78 My model of St Moritz had a track length of 1,722 m with a gradient of 8.1°. I used the friction and drag characteristics from Norbert Grün's equations above.

Chapter 8: A leap into the unknown

79 Kidwell, C.B., *Women's Bathing and Swimming Costume in the United States* (Smithsonian Institute Press, 1968). Available as an eBook.

80 Only men competed in swimming at the first Olympics; this might have been a demonstration event. http://www.onlinefootage. tv/stock-video-footage/7895/1st-olympics-athens-1896-female-swimmer?keywords.

81 Gibson, E. and Firth, B., *The Original Million Dollar Mermaid: The Annette Kellerman Story* (Crows Nest, Australia, Allen and Unwin, 2005).

82 Raszeja, V.M., 'Clara Dennis'. *Australian Dictionary of Biography* (National Centre of Biography, Australian National University, 1993) http://adb.anu.edu.au/biography/dennis-clara-clare-9951/text17629.

83 Sam Knight (2008), 'The tragic story of Wallace Hume Carothers', *The Financial Times*, 29 November 2008. https://www.ft.com/content/2eae82b2-b9fa-11dd-8c07-0000779fd18c?mhq5j=e1.

84 Voyce, J., Dafniotis, P. and Towlson, S., 'Elastic Textiles' in *Textiles in Sport*, R. Shishoo (ed.) (Cambridge, Woodhead Publishing, 2005).

85 Davies, E. (1997), 'Engineering Swimwear', *The Journal of the Textile Institute* 88(3), pp. 32–36.

86 European Patent Office (2009), 'A revolutionary swimsuit'. http://www.epo.org/learning-events/european-inventor/finalists/2009/fairhurst.html.

87 Kessel, A. (2008), 'Born Slippy', *The Guardian*, 23 Nov 2008. https://www.theguardian.com/sport/2008/nov/23/swimming-olympics2008.

88 Neiva, H.P., Vilas-Boas, J.P., Barbosa, T.M., Silva, A.J. and Marinho, D.A. (2011), 'Analysis of swimsuits used by elite male swimmers', *Journal of Human Sport & Exercise* 6(1), pp. 87–93.

89 Abrahams, M., (2006), 'Skinny Dipping like Dolphins', *The Guardian*, 6 June 2006. www.theguardian.com/education/2006/jun/06/research.highereducation.

90 Toussaint, H.M., Truijens, M., Elzinga, M-J., Van de Ven, A., De Best, H., Snabel, B. & De Groot, G. (2002), 'Effect of a Fast-skin™ "Body" Suit on Drag during Front Crawl Swimming', *Sports Biomechanics* 1(1), pp. 1–10.

[91] Dean, B. and Bhushan, B. (2010), 'Shark-skin surfaces for fluid-drag reduction in turbulent flow: a review', *Phil. Trans. R. Soc. A* 368, pp. 4775–4806.

[92] Kainuma, E., Watanabe, M., Tomiyama-Miyaji, C., Inoue, M., Kuwano, Y., Ren, H., Abo, T (2009), 'Proposal of alternative mechanism responsible for the function of high-speed swimsuits', *Biomed Research* 30(1), pp. 69–70.

[93] http://www.theage.com.au/news/sport/swimming/suit-worth-two-seconds-in-germans-record-swim/2009/07/27/1248546678468.html.

[94] https://engineeringsport.co.uk/2011/09/18/swimsuit-ban-will-affect-world-record-progression/.

Chapter 9: Framing the problem

[95] The fact that the air density is lower when the humidity is higher surprises most people because the air tends to feel thicker and heavier. However, it can be easily explained by a simple thought experiment. Air is 78% nitrogen and 21% oxygen with a little bit of argon and a few other minor components such as methane and hydrogen. Nitrogen and oxygen are present as N_2 and O_2 which have molar masses of 28 and 32 grams per mole respectively. The mix of all these components gives air a molar mass of about 29 grams per mole. Now consider replacing some of this with water vapour which is made up of H_2O molecules. Hydrogen is the lightest substance with a molar mass of 1 gram per mole and H_2O has a molar mass of only 18 grams per mole so that removing any air and replacing it with water vapour would have to decrease the density of the air. https://www.engineeringtoolbox.com/molecular-mass-air-d_679.html.

96 De Coubertin, P., Philemon, T.J., Politis, N.G. and Anninos, C., *The Olympic Games B.C.–A.D. 1896* (London, H. Grevel and Co., 1897). The official record describes six competitors, although some say there were seven. The only contemporary picture I could find actually had eight bikes on the start line, but one might have been a pacemaker, official or coach. http://library.la84.org/6oic/OfficialReports/1896/1896part2.pdf.

97 There is no definitive source to this saying but a popular story has it that it was said by English bowler Cliff Gladwin. The year was 1948 and England were playing South Africa; Gladwin came in to bat with 12 runs needed off the last three overs. He got a leg bye with the last ball to win the match. Afterwards in the dressing room he showed off the bruise from where it hit his leg and was quoted as saying, "I told you, 'cometh the hour, cometh the man'."

98 Before all the engineers reading this shout at me, a strain gauge actually measures strain rather than deflection. Strain is the normalised deflection δ compared to its original length L and calculated using $\varepsilon = \delta / L$; for example, if the gauge was five millimetres long and it lengthened an additional millimetre when the pedals were pushed, the strain would be 20 per cent.

99 The power crank measures the rate of rotation of the crank ω in radians per second and calculates power in watts using the equation $P = F\omega$, where F is the force determined using the strain gauges.

100 Wolfgang Menn has a good website with lots of data and shows his sources. However, it's difficult to tell where the original data came from. www.wolfgang-menn.de/hourrec.htm.

101 Bassett, D.R. Jr., Kyle, C.R., Passfield, L., Broker, J.P and Burke, R. (1999), 'Comparing cycling world hour records, 1967–1996: modelling with empirical data', *Med. Sci. Sports Exerc.* 31(11), pp. 1665–1676.

[102] Burke, E.R., *High Tech Cycling* (Champaign, IL, Human Kinetics, 2003).

[103] *Strava* puts everything you would expect into an estimate of cycling power; they include rolling resistance, aerodynamic drag and gravity in their equation. https://support.strava.com/hc/en-us/articles/216917107-Power-Calculations.

Chapter 10: The flashing blade

[104] Hines, J.R., *Figure Skating: A History* (Urbana and Chicago, University of Illinois Press, 2006), p. 18.

[105] Versluis, C. (2005), 'Innovations on thin ice', *Technovation* 25, pp. 1183–1192.

[106] If either of the surfaces deform significantly, then the coefficient of friction can be greater than one. Often this is referred to as the coefficient of traction.

[107] De Koning, J.J. (2010), 'World Records: How much athlete? How much technology?', *International Journal of Sports Physiology and Performance* 5, pp. 262–267.

[108] Van Ingen Schenau, G.J. (1982), 'The influence of air friction in speed skating', J. *Biomechanics* 15(6), pp. 449–458.

[109] Saetran, L., Oggiano, L., 'Skin Suit Aerodynamics in Speed Skating' in Nørstrud, H. (ed.) *Sport Aerodynamics* (Springer, Vienna, CISM International Centre for Mechanical Sciences, vol. 506, 2008), pp. 93–105.

[110] Christine Nesbitt's record was 1 minute 12.68 seconds, set in Calgary on 28 January 2012.

[111] De Koning, J.J. (1997), 'Slapskate history and background', 20 Feb 1997. www.sportsci.org/news/news9703/slapxtra.htm.

[112] Van Ingen Schenau, G.J., De Koning, J.J., De Groot, G., Scheurs, A. W. and Meester, H. (1996), 'A new skate allowing powerful flexions improves performance', *Medicine and Science in Sport and Exercise* 28(4), pp. 531–535.

[113] De Koning, J.J., Houdijk, H., De Groot, G. and Bobbert, M.F. (2000), 'From biomechanical theory to application in top sports: the Klapskate story', *Biomechanics* 33, pp. 1225–1229.

[114] Foster, L. (2009), 'The effect of technology on elite performance', PhD thesis, Sheffield Hallam University, Sheffield, UK.

Chapter 11: Superheroes

[115] Terry Willett was injured in a mining accident in 1963 and transferred to Lodge Moor specialist unit on the edge of Sheffield. He used sport as part of his rehabilitation, going on to win medals in basketball and his favourite sport, fencing. You can read his interview here: www.paralympicheritage.org.uk/terry-willett.

[116] Frankel, L., Michaelis, L.S., Golding, D.R. and Beral, V. (1972), 'The blood pressure in paraplegia', *Paraplegia* 10, pp. 193–198. www.nature.com/articles/sc197232.pdf?origin=ppub.

[117] The women's 800 metres demonstration event was held in Los Angeles, 1984. The commentators do their best to be politically correct with only a few slip-ups. www.youtube.com/watch?v=LrgzK3NWVbs.

[118] The women's 800 metres demonstration event in Seoul, 1984 for the 'physically challenged'. www.youtube.com/watch?v=WJt30sBM-Eg&t=36s.

119 Cooper, R.A. (1990), 'Wheelchair racing sports science: a review', *Journal of Rehabilitation Research and Development* 27(3), pp. 295–312.

120 The project was a collaboration between the two Sheffield Universities and was led by Prof Edward Winter from Sheffield Hallam University. My team and I subsequently moved there in 2006.

121 Van Phillips tells his story here: https://www.youtube.com/watch?v=4d7O8UxFJt4.

122 Nolan, L. (2008), 'Carbon fibre prostheses and running amputees: a review', *Foot & Ankle Surgery* 14, pp. 125–129.

123 Pistorius, O., *Blade Runner: My Story* (London, Virgin Books, 2009). For example, Chapter 4 is titled 'Carpe Diem' (Seize the Day) and begins, "When the time came for me to begin high school, my parents, true to form, gave me free rein."

124 Ross Tucker and Jonathan Dugas write a great blog called 'The Science of Sport'. They give detailed accounts of the science behind many of the controversies of the day, including that of Oscar Pistorius. Have a read – you'll find that a day has suddenly passed by. http://sportsscientists.com/thread/oscar-pistorius/.

125 Pickup, O., (2012), 'London 2012 Olympics: Games legend Michael Johnson believes Oscar Pistorius has an "unfair advantage"'. *Daily Telegraph*, 17 July 2012 http://bit.ly/2Gdkmke.

126 There are three ways to work out the deflection of the blades and the subsequent stresses in the material: the first way is to use strain gauges attached to the surface; the second is to use a computer technique called Finite Element Analysis; the third way is a back-of-the-envelope analysis of the structural dynamics of the blade,

as follows. Consider the prosthesis as a simple straight cantilever beam (like a diving board but vertical) of length L and rectangular cross section of breadth b and depth d. The fixed end would be at the stump and the free end on the ground. During contact with the ground, the runner applies a horizontal force F at the free end to deflect it backwards by an amount δ. The relationship between the force and deflection at the end of a cantilever is given by $F=(3EI/L^3)\delta$ where E is the Young's modulus of the material and $I=bd^3/12$. The part in brackets is effectively the stiffness of the blade when deflected at its end. The designers will play around with the stiffness so that the blade deflects just the right amount for the runner. If the length increased by 2.25%, as it did with Oliveira, then the stiffness would have decreased by the cube of this, i.e. 7% and the deflection would have gone up by the same amount during stance. If this was excessive, then they could compensate by increasing the number of carbon fibres (thus increasing E by 7%), or the breadth (increasing b by 7%), or the depth (increasing d by 2.25%). This gives an idea about how to improve design (although one should remember that a J-shaped blade isn't a straight cantilever and so the answers are only approximate).

127 Oliveira got even better, going on to break the world records in 2013 in the 100, 200 and 400 metres.

128 This is by volume; by weight it is 50 per cent. This shows how efficient the use of carbon is.

Chapter 12: Sports technology in the new millennium

129 Lewis, M., *Moneyball: the Art of Winning an Unfair Game* (New York, W.W. Norton and Company, 2004).

130 The UK Sport Innovation Partnership is now administered by the English Institute of Sport under the creative guidance of Gavin Atkins.

131 Westenberg, T., 'Timing light accuracy in sport' in *The Engineering of Sport*, Haake, S.J. (ed.) (Oxford, Blackwell Science, 1998), pp. 291–299. Westenberg worked for the United States Olympic Committee in Colorado Springs so knew his stuff. He explained that the pencil beam of a set of timing lights was switched on and off at 1,000 times per second. When a runner passed through them, it could take up to ten pulses for the light intensity to drop to zero: this took one hundredth of a second.

132 Leger, L. and Thivierge, M. (1988), 'Heart rate monitors: validity, stability and functionality', *Physician and Sports Medicine* 16, pp. 143–151. It's a little perplexing how the heart rate monitors systematically measured heart rate too low. You would expect values from the less accurate heart rate monitors to be scattered about the values from the gold-standard ECG method, but instead they measure up to 20 beats per minute lower. Leger suggested this was something to do with the algorithms used by the different manufacturers – unfortunately these remain commercial secrets.

133 Roth S.J. (2006), 'Why does lactic acid build up in muscles?' *Scientific American*, 23 January 2006. www.scientificamerican.com/article/why-does-lactic-acid-buil/.

134 Jones, A.M. (2006), 'The physiology of the world record holder for the women's marathon', *International Journal of Sports Science and Coaching* 1(2), pp. 101–116.

135 BBC (2016), 'Greater Manchester Marathon course was 380m short', says measuring body, 21 April 2016 http://www.bbc.co.uk/sport/athletics/36104638.

[136] Trilateration uses the coincidence of spheres to identify 3D location rather than triangles and angles as used in triangulation. One question often asked is why four satellites are needed rather than three to identify 3D position. With one satellite, all we know is that the position of a person on the earth lies somewhere on the surface of a digital sphere with the satellite at its centre. If there is a second satellite, then there is a second sphere and the person is somewhere on the lines where the two spheres intersect; these are two arcs shaped together to make a pair of lips. The sphere of the third satellite and sphere intersects with this and identifies two points, one on each lip, while the fourth satellite identifies which point is the correct one.

[137] Aughey, R.J. (2011), 'Applications of GPS technologies to field sports', *International Journal of Sports Physiology and Performance 2011*, 6, pp. 295–310. Many researchers used early GPS without understanding its limitations. For short sprints of ten metres or so, errors could be as high as 30 per cent. Aughey says, "It is unlikely that studies conducted in team sport matches with 1 Hz GPS can detect anything other than total distance moved by players."

[138] Anderson, C. and Sally, D., *The Numbers Game: Why Everything You Know About Football is Wrong* (London, Viking, 2013), p. 45. The original data comes from Infostrada. A distribution satisfies Benford's law if the probability $P(d)$ of a leading digit d occurring in a set of numbers is given by $P(d) = log_{10} (1+1/d)$. For example, when $d=1$, $P(d) = log_{10} (2) = 0.301$ and the digit 1 is expected to lead 30.1% of the time. In the Premier League in 2011–12, it seems that the number of passes made was not far off Benford's law.

[139] Slot, O., *The Talent Lab* (London, Ebury Press, 2017). Slot interviewed Simon Timson and Chelsea Warr, the performance directors of UK Sport during the successful years of London 2012 and Rio 2016. Slot describes the UK Sport Medal Tracker which has 200 names of individuals and their chances of getting a medal.

This focused the system on the task at hand: winning medals. The data we collected for the sports we worked with turn into simple performance metrics appropriate for the sport which allowed Timson and Warr to assess their medal chances and intervene if something seemed to be going wrong. Our estimates are that our work supported 24 medals in London and 42 medals in Rio.

Chapter 13: The next big thing

[140] Cross, R. and Bower, R. (2006), 'Effects of swingweight on swing speed and racket power', *Journal of Sports Sciences*, 24(1), pp. 23 30. The authors showed that there is a broad range of racket masses of about five to eight times the mass of the ball, where players get the maximum ball speed when they hit it. As the racket mass decreases to zero, so does the ball rebound speed: effectively, the incoming ball would knock the racket out of the way.

[141] Lee, C., Linkenauger, S.A., Bakdash, J.Z., Joy-Gaba, J.A., Profitt, D.R. (2011), 'Putting Like a Pro: The role of positive contagion in golf performance and perception', *PLoS ONE* 6(10), e26016.

[142] Bérdi, M., Köteles, F., Szabó, A. and Bárdos, G. (2011), 'Placebo effects in sport and exercise: a meta-analysis', *European Journal of Mental Health* 6, pp. 196–212.

[143] Porcari, J., Otto, J., Felker, H. et al. (2006), 'The placebo effect on exercise performance [abstract]', J. *Cardiopulmonary Rehabilitation and Prevention* 26(4), p. 269. Discussed in: Beedie, C.J. and Foad, A.J., (2009), 'The placebo effect in sports performance: a brief review', *Sports Medicine* 39(4), ProQuest, p. 320.

[144] Reardon, S. (2016), '"Brain doping" may improve athletes' performance', *Nature* 531, pp. 283–284, 17 March 2016.

[145] Okano, A.H, Fontes, E.B., Montenegro, R.A. et al. (2015), 'Brain stimulation modulates the autonomic nervous system, rating of perceived exertion and performance during maximal exercise', *British Journal of Sports Medicine* 49, pp. 1213–1218.

[146] Edwards, D.J., Cortes, M., Wortman-Jutt, S., Putrino, D., Bikson, M., Thickbroom, G. and Pascual-Leone, A. (2017), 'Transcranial direct current stimulation and sports performance', *Frontiers of Human Neuroscience* 11, p. 243.

[147] Haake, S., James, D. and Foster, L. (2015), 'An improvement index to quantify the evolution of performance in field events', *Journal of Sports Sciences* 33(3), pp. 255–267. The paper took accessible performance data such as times in running and distance in field events and used them to estimate energy expended in doing the event. For example, heights in the high jump allow potential energy to be estimated and comparison of two performances for an athlete of the same height and weight gives a ratio of the two heights, h_2/h_1, where jump of height h_2 is compared to jump of height h_1. For instance, the average of the top 25 female high jumpers in 2012 jumped 1.966 metres compared to their peers of 1948 who jumped 1.591 metres; this gave a performance improvement of 23.5%.

[148] R&A and USGA (2017) 2017 Distance report. www.randa.org/News/2018/03/Distance-Report.

[149] Haake, S.J, James, D.M. and Foster, L.I. (2014), 'An improvement index to quantify the evolution of performance in running', *Journal of Sports Sciences* 32:7, pp. 610–622. The performance improvement index was calculated using the same approach as Haake *et al.* above; in running the dominant energy expenditure is that of aerodynamic drag. This means that if two running times are compared, then the ratio t_1^2 / t_2^2 is used, where run of time t_2 is compared to run of time t_1. For example, the average of the top 25 female 100 metre runners in 1948 ran a time

of 12.06 seconds compared to their peers of 2012 who ran in a time of 10.98 seconds; this gave a performance improvement of 20.6%.

[150] National Research Council of the National Academies (2012), Human Performance Modification: Review of Worldwide Research with a View to the Future, The National Academies Press, Washington. Available from: http://nap.edu/13480.

[151] Wang, Y.X., Zhang, C.L., Yu, R.T., Cho, H.K., Nelson, M.C. et al. (2004), 'Regulation of muscle fibre type and running endurance by PPARδ', PLoS Biol 2(10), e294.

[152] Hanson, R.W. and Hakimi, P. (2008), 'Born to run; the story of the PEPCK-C^{mus} mouse', Biochimie 90(6), pp. 838–842.

[153] Deutsche Welle (2006), 'German Coach Suspected of Genetic Doping', 3 February 2006. http://bit.ly/2GmfucF.

[154] NBC (2008), 'China caught offering gene doping to athletes', 23 July 2008. http://nbcnews.to/2txXiKk.

[155] Butler, N. (2017), 'IOC President not convinced e-sports reflects Olympic rules and values', Inside the Games, 25 April 2017. http://bit.ly/2FJiodr.

[156] D'Arcy, K. (2016), 'Athlete Michael Rix: I won a triathlon 12 weeks after my hip implant', Daily Express, 19 July 2016. http://bit.ly/2FuJPbJ.

[157] One theory about the phrase 'back to square one' is that it came from the Radio Times which carries listings for radio and TV. It had an image of the pitch split up into eight squares which were used during radio commentary to help listeners understand the position of the players. The phrase 'back to square one' was retrospectively

said to have identified that the ball was going back to the keeper. This can't be right as the keeper wasn't always in square one and could have been at the other end in square eight. It's more likely that the phrase originated with the rise of board games in the 1950s.